记忆空间
Remember
in Architecture

建筑立场系列丛书 No.88

[德]里伯斯金建筑事务所 等 | 编

于凤军 钱进 罗茜 王晴 刘欣 王方冰 | 译

大连理工大学出版社

记忆空间

004　城市与记忆_Olga Sezneva

014　记忆之家_baukuh

024　阿尔托大学主楼Dipoli大楼_ALA Architects

036　海法大学Younes & Soraya Nazarian图书馆_A. Lerman Architects

048　少女塔的修复_De Smet Vermeulen Architecten + Studio Roma

058　安吉·雷迪博士纪念园_Mindspace

070　Imnang 文化公园_BCHO Architects

084　Len Lye 中心_Pattersons

094　记忆并非完美_Diego Terna

104　于特岛赫恩胡塞特纪念馆和学习中心_Blakstad Haffner Architects

120　科泰艺术展馆_SANDWICH

132　马科瓦的Ulma家庭博物馆_Nizio Design International

140　卡廷博物馆_BBGK Architekci

154　大地的见证——侵华日军第七三一部队罪证陈列馆
　　_Architectural Design & Research Institute of SCUT South China University of Technology

168　赫茨尔山阵亡士兵纪念馆_Kimmel Eshkolot Architects

186　渥太华国家大屠杀纪念碑_Studio Libeskind

200　第二次世界大战博物馆_Studio Architektoniczne Kwadrat

216　建筑师索引

C3 建筑立场系列丛书 No.88

Memory in Architecture

004 City and Memory _ Olga Sezneva

014 House of Memory _ baukuh

024 Dipoli, Aalto University Main Building _ ALA Architects

036 University of Haifa, Younes & Soraya Nazarian Library _ A. Lerman Architects

048 Restoration of the Maiden Tower _ De Smet Vermeulen Architecten + Studio Roma

058 Dr. Anji Reddy Memorial _ Mindspace

070 Imnang Culture Park _ BCHO Architects

084 Len Lye Center _ Pattersons

094 Memory's not Perfect _ Diego Terna

104 Hegnhuset, Memorial and Learning Center in Utøya _ Blakstad Haffner Architects

120 KOHTEI _ SANDWICH

132 The Ulma Family Museum in Markowa _ Nizio Design International

140 Katyń Museum _ BBGK Architekci

154 The Witness of Land – Memorial Hall of Crime Evidences by Unit 731
_ Architectural Design & Research Institute of SCUT South China University of Technology

168 Mount Herzl Memorial Hall _ Kimmel Eshkolot Architects

186 National Holocaust Monument Ottawa _ Studio Libeskind

200 The Second World War Museum _ Studio Architektoniczne Kwadrat

216 Index

记忆空间

Mem
in Archi

城市与记忆 / **Olga Sezneva**
记忆并非完美 / **Diego Terna**

nory
tecture

城市与记忆
City and Memory

Olga Sezneva

  如果说雕像就是为了纪念某个人或某起事件而修建的（如牛津字典所定义的那样），那么罗伯特·李将军就有麻烦了。现在，人们对这位将军英雄事迹的记忆永远被一位年轻女性希瑟·海耶的死亡蒙上了一层阴影。2017年夏天，美国弗吉尼亚州夏洛茨维尔市爆发了一场抗议活动，随后抗议发展成了暴乱，希瑟·海耶在这起事件中丧生。不过，罗伯特·李将军的麻烦并不是2017年夏天才开始的。早在2015年，李将军以及其他一些南部邦联人士就成为公众质疑的对象。在路易斯安那州新奥尔良、佛罗里达州的盖恩斯维尔或杰克逊维尔等城市，他们的雕像都被拆掉了。对南部邦联英雄人物雕像进行批判性审查的潮流后又蔓延到美国北部并不断扩大。在纽约市，许多雕像也遭到人们的非议。一些要求被拆除的雕像包括：矗立在自然历史博物馆之外的西奥多·罗斯福总统的骑马雕像，位于第五大道和103街上的妇科医生J.马里恩·辛斯的雕像，还有位于中央公园西南角的克里斯托弗·哥伦布纪念碑。现在，铭刻在人们记忆里的是：罗斯福倡导优生学，辛斯凭借对南方黑人女性不使用麻醉药进行医学实验而获得他作为"妇科手术之父"的美誉，而哥伦布发现新大陆导致了土著人的灭绝。

  2007年，在地球另一端的波罗的海沿岸，相关部门计划拆除一座纪念碑，但却遭到了这个城市里近一半人口的反对和抗议。一时之间，纪念碑成为公众热议的对象。这座城市便是塔林，被称作"青铜战士"的纪念碑早在半个世纪之前就赫然矗立在市中心，十分显眼。此处的纪念碑与李将军雕像不同，它不是用来纪念任何具体的人的，它代表的是整个国家，尤其是国家的军事力量：苏联红军。尽

If the memorial is indeed a sculpture established to remind of a person or an event (as the Oxford dictionary tells us), then General Robert Lee is in trouble. The memory of his deeds is now forever overshadowed by that of the death of a young woman, Heather Heyer, who was killed in Charlottesville, VA, USA in the summer of 2017 when a protest turned violent. The General's trouble has not started there, however. He, as well as a half-a-dozen of other Confederate figures, fell under public scrutiny earlier, in 2015; their statues were taken down in cities like New Orleans, LA, Gainesville or Jacksonville, FL. The tide of critical examination of the Confederate memorials then moved north and widened, reaching the city of New York and bringing its own statuary to question. Monuments threatened with removal included the equestrian statue of President Theodore Roosevelt outside the Museum of Natural History and the statue of the gynecologist J. Marion Sims on Fifth Avenue and 103rd Street, in addition to the memorial of Christopher Columbus at the southwest corner of Central Park. It has been now remembered that Roosevelt advocated eugenics and Sims gained his reputation as a pioneering gynaecologist by performing medical experiments on black women in the South without the use of anaesthesia[1], while the discoveries of Columbus led to the extermination of indigenous people.

On the other side of the globe, near Baltic shore, in 2007, half of the population of a city rebelled when plans were made to remove a memorial, making monuments a popular object of public contention. The city was Tallinn and the memorial called the Bronze Soldier was erected prominently in the city center half-a-century earlier. The memorial was different from that to General Lee. It did not commemorate any concrete individual, but represented a nation, and specifically, its military force: the Soviet. Still, it stirred controversy decades after it was constructed for a

希瑟·海耶纪念地,美国弗吉尼亚州夏洛茨维尔第四街道西南
Memorial for Heather Heyer on 4th Street SE in Charlottesville, Virginia, USA

罗伯特·爱德华·李将军雕像,美国弗吉尼亚州夏洛茨维尔
Robert Edward Lee Sculpture in Charlottesville, Virginia, USA

管如此,在其建立数十年后还是因为类似的原因(社会风气与国际政治格局已经发生了变化)出现了争议。随着苏联解体,爱沙尼亚成为欧盟与北约成员国,矗立的士兵雕塑却代表着苏联和俄罗斯人对其的占领和统治,因此占这个国家主体的爱沙尼亚族人希望将其拆除,而占少数的俄罗斯族人——其中几乎近一半的人非爱沙尼亚公民身份,仅享有居住权——则对此持反对意见。因为于他们而言,拆除纪念碑意味着将他们自身从历史中抹去。因此,他们发动暴乱(造成一人死亡),对爱沙尼亚工商企业发动网络攻击,最后迫使爱沙尼亚政府做出妥协:不拆毁纪念碑,但将其从市中心迁移到该市更加庄严肃穆的公墓。

纪念碑或雕像并不总是与暴力联系在一起的。大多数时候,这些纪念碑或雕像仅仅在城市景观中起装饰性作用,除了鸟儿之外,人们几乎很少注意到它们,只有在举行为数不多的正式庆典的时候,它们才会进入社区的公共生活当中。在这一点上,它们无一例外:大多数纪念碑或雕像都是以这样的方式进入人们生活的。没有一块大理石或锻铁本身就蕴含着历史;对于具有纪念意义的纪念碑而言,它需要嵌入当前关系与当代政治进程的网络中,需要与其他的文化符号展开对话。

但是这一切与城市又有什么关系呢?城市为制造这样的记忆提供了材料、空间以及专门知识。理查德·桑内特曾写道:"……为了认同一个人,你认同了一个空间",[2] 而城市中恰恰满是这样的空间。这些空间拥有自然的地形,成为文化记忆深植的土壤,或者说,这些地方成为生活"素材"的档案库。这一点在路街的名字和地方的名字上表现得最为明显,但又不仅仅局限于此。过去的影子在建筑与

similar reason: societal mores and international politics have changed. With the decomposition of the communist block and Estonia choosing the route of membership in the European Union and the NATO, the soldier stood to represent the Soviet and the Russian occupation. The ethnic Estonian majority wanted it to be removed. The ethnic Russian minority – nearly half of it not carrying Estonian citizenship but a mere residency – resisted. To them, the removal of the monument signalled their own erasure from history. They rioted leaving behind one dead, launched cyberattacks on Estonian businesses, and eventually forced a compromise: the memorial was not demolished but moved from the center to the city's more venerated cemetery.

Violence has not always been part of the memorials' lives. Most of the time they stood as mere decorative additions to townscapes hardly noticed by all but birds, entering public lives of communities at rare moments of official celebrations. In that, they were not an exception: most memorials do that. No piece of marble or wrought iron contains history in itself; for a memorial to possess a mnemonic power, it needs to be inserted in a web of current relationships, contemporary political processes, and enter a dialogue with other cultural symbols.

However, where does this all have to do with the city? Cities provide the material, the space and the expertise with which such memories are made. Richard Sennett once wrote that, "…you identify with a space in order to identify with another human being"[2], and cities are filled with such spaces. They are physical terrains in which cultural memory is embedded – or even better, where the "stuff" of life is archived. This is most evident in the names that streets and places bear, but it is not all. The presence of the past is visible in the architecture and the street layout, the offices and the shops, the bars and the cafes, and even the greenery. But which history? In the body of the

克里斯托弗·哥伦布雕像，美国纽约
statue of Christopher Columbus, NYC, USA

军事墓地的青铜战士雕像，爱沙尼亚塔林
Bronze Soldier at the Military Cemetery in Tallinn, Estonia

街道布局、办公室与商店、酒吧与咖啡馆中随处可见，甚至在绿色植物中也能追寻到一些它的踪迹。但是，人们铭记的又是怎样的历史呢？巴黎著名的纪念性建筑——圣心大教堂记录的全是法国共和主义阶级斗争的历史，记录的是巴黎公社的历史。³其漫长的建造史（译者注：历时44年）及其尴尬的外形（译者注：兼具拜占庭建筑风格和罗马建筑表现手法）都反映了长达一个世纪以来的压迫与反抗、资本主义的成熟和民族自豪感。另一个例子便是柏林的共和国宫。共和国宫在21世纪初被迫拆除，为重建霍亨索伦王朝的宫殿让位。共和国宫被迫拆除是因为该建筑的外形和给人的感觉以及建造共和国宫所使用的混凝土的颜色。⁴在如今隶属俄罗斯的加里宁格勒市，经历过第二次世界大战和苏联时期世界格局重组洗礼的树木都能让人看到德国柯尼斯堡（译者注：与加里宁格勒同音同义）市的影子，德国柯尼斯堡现在已经不存在了，是加里宁格勒的前身。⁵这儿要提到的最后一个例子让人们想起了一个世纪以前李格尔（译者注：Alois Riegl, 1858—1905年，19世纪末20世纪初奥地利著名艺术史家）所指出的精心打造的纪念性建筑和无心为之的纪念性建筑之间的区别：一些遗址、建筑物，甚至它们的遗迹都会因为一个社会的意志成为人们的记忆，而不是设计师的想法。⁶诚然，任何事物都可以充满记忆的力量。Urville、Azeville以及Biville都是诺曼底地区的小型社区，它们因为摄影艺术家Jane Wilson和Louise Wilson所拍摄的第二次世界大战时诺曼底防线遗迹而举世闻名。一封信（一张纸），一幅画（一张图、一张照片、一份数字文件）以及一朵干枯玫瑰都蕴含着一起事件的历史、一个废墟的历史、一个遗迹的历史，唤起人们对此的记忆，它们可能会成为历史的载体。

famous Paris monument, Basilique du Sacré-Cœur, is written an entire history of class struggle, the French Republicanism and of the Paris Commune.³ Its long coming to existence, as much as its awkward shape reflects a long century of oppression and resistance, the maturing of capitalism and nationalistic pride. Another example is Palast der Republik in Berlin, which had to be demolished in the early 2000s to give way to a replica of Hohenzollern Palace, all because of the shape and the feel of the building, and the colour of the concrete of which it was made.⁴ And in the Russian city of Kaliningrad which nowadays belongs to Russia, trees that survived WWII and the Soviet-period reorganization stand to be living witnesses of the German Königsberg, a non-existent city that Kaliningrad once was.⁵ This last example brings to mind the distinction made by Alois Riegl a century ago between the intentional and unintentional monuments: some sites, buildings and even their traces turn into memory topoi by the will of a society, not intentions of their designers.⁶ Truly anything may become imbued with mnemonic power. Urville, Azeville and Biville, small communities in Normandy, became internationally known when ruins of the WWII defence lines attracted photography artists Jane Wilson and Louise Wilson. The way a letter (a piece of paper), a picture (a drawing, a photograph, a digital file), and a dry rose together summon and contain a history of an affair, a ruin, a trace, may be vectors to history.

The city throws up its own mnemonic effect, as literary creations from Walter Benjamin's "Moscow Diary", to Stratis Tsirkas' trilogy "Drifting Cities", to the recent Teju Cole's "Open City", poignantly show. The city has its own way of connecting cultures, generations, technologies, the domains that are human and nonhuman, even spectral. The infamous Hungry Ghost festival in Singapore is one example of the latter. In its course, paper money, paper houses,

霍亨索伦宫，柏林，1950年
Hohenzollern Palace in Berlin, 1950

共和国宫，柏林，于2008年拆除
Palace of the Republic, Berlin, before the demolition in 2008

　　城市本身能唤起人们许多记忆，如瓦尔特·本雅明的《莫斯科日记》、Stratis Tsirkas的三部曲《漂流的城市》以及最近出版的泰茹·科尔的《开放城市》等文学作品所展示的那样。城市有其自己的方式将文化、世代传承、技术以及各个领域（人类、非人类甚至妖魔鬼怪）都联系在一起。关于鬼怪之说，"远近闻名"的新加坡"饿鬼节"（译者注：中元节）就是其中一个。节日期间，人们在有些地方的人行道边、街头拐角、车库外以及美食广场等地焚烧纸钱、纸房子，甚至纸做的宝马车，因为直到20世纪90年代城区重新规划改造之前，这些地方都是城市里华人安葬死者的"坟场"。为了慰藉死者，生者继续为他们带来祭品。当夜幕降临，"第二张地图，即历史上幽灵们居住的地方，出现在人们最熟悉的地方，这是人们记忆与所熟悉的地形地貌的彻底分离……出现在一切都受到超级控制的当代城市之中"[7]。这就是"城市就是一个档案馆"这一隐喻的含义。这一隐喻非常恰当，城市里的许多建筑环境都以其体现的形式捕捉生活中的细枝末节，并作为想象历史的源泉得以流传。建筑环境是一个集合体，集艺术品、活动的沉淀模式以及实践于一体，充分体现在建筑构造甚至其所在场地上，成为标志，成为象征，蕴涵着历史，记录着过去，仍然与现在琴瑟共鸣，是另一种记录和保存的方法。

　　新加坡"鬼怪"的存在也显示了某些其他重要的东西。在这里，我们不仅谈论分类，还可以谈论形成记忆（和失忆）的不同方式；其中传递的不仅是思想与事实，还有经历与情感；在唤醒过去的过程中不同的生命也得以被召唤。这便引出另一种思考：每个空间都有展示过去、为过去发声或使过去沉默的不同方式。街道、人行道、铺设的路面这些构成典型城市生活的元素都是记忆生态的组成部分。

even paper BMWs are burnt on sidewalk curbs, street corners, parking garages and food courts in the areas, which until the urban renewal of 1990s were burial sites for the city's Chinese community. To tame the dead, the living continue to bring offerings. As darkness falls, "a second map, a ghostly historical topography, appears on top of the familiar one, a radical disjuncture of memory and topography... within the hyper-controlled surfaces of the contemporary city"[7]. This is "the city as an archive" metaphor, which works well because so much of the city's built environment captures the minutia of life in its embodied form, circulating as a source of historical imagination. It is an assemblage of artefacts, sedimented patterns of activities and practices embedded in its fabric and even its ground, as signs and representations, as a collection and a record of the past that continues to resonate in the present. It is another method of filing and preserving.

The spectral presences of Singapore bring out something else which is important. Here, not only of cataloging, we may also speak of different ways in which memory (and amnesia) are "done", in which what are transmitted are not only thoughts and facts, but also experiences and emotions, and in which different beings are summoned in the processes of evoking the past. This leads to a different kind of consideration: each space has different ways of showing, articulating or silencing the past. A street, a sidewalk, a pavement – the elements of a typical urban life – are parts of an ecology of memory.

Much of the statuary in the modern city works by appealing to contemplation. Strategically placed, monuments are the object of viewing. They participate in urban spectacle. They create perspectives and vistas. Scaled to impress, they commonly appeal to collectivities. The Berlin-born artist Gunter Demnig recognized a different possibil-

哥尼斯堡塔楼，俄罗斯加里宁格勒
Königsberg Castle tower, Kaliningrad, Russia

海滩与房屋遗迹，法国Urville-Nacqueville
Beach and Vestiges of Blockhouse, Urville-Nacqueville, France

现代城市中的许多雕塑都是通过引发人们沉思而发挥自身作用的。从其选址的策略来看，纪念性建筑是人们瞻仰的对象，也是城市景观的一部分，为人们了解城市提供了不同的视角，使人们对未来充满希望。这些纪念性建筑通常都非常宏伟，能够吸引公众的目光。出生于柏林的艺术家冈特·德姆尼希发现城市空间中存在着一种不同的可能性，即城市空间特有的交际功能。他发明了一种通过安装记忆触发器而不是通过设立雕像的方式来使人们回忆过去，铭记历史。自20世纪90年代中期开始，他将"绊脚石"铺设在柏林街头一些房屋前的人行道上。这些房屋正是犹太人受害者被赶往集中营之前的住所。最近，该项目扩展到纪念其他纳粹受害者们，其中包括：罗姆人、同性恋者和残疾人。"绊脚石"的大小和形状与常见的铺设路面的方砖无异，表面的黄铜上铭刻有受害者的姓名、生前住址以及死亡的细节。这些逝者的名字没有被安放在特定的地点，而是与马路融为一体。人们行走其上，会注意到它们的存在。这一想法与犹太人必须在一年一度的逾越节家宴上宣读《哈加达》（译者注：哈加达是希伯来文的名词，意思是"告诉"。犹太人的父母需在逾越节向子女讲述当年出埃及的故事，突出庆祝逾越节的主题：对获得自由的一种庆祝。）故事以纪念以色列人走出埃及的做法不谋而合。《哈加达》要求信徒不要记住历史事件，而是将自己置身于事件发生时，置身于神出现的那一刻，置身于走出埃及的那一刻（不是过去人们如何走出埃及），因此，从某种程度上来说，是要重现那一场景[8]；每一代人都要把自己想象成是从埃及逃出来的。柏林街头每天都在上演的行为，即人们每天都要经过那些"绊脚石"，也是同样的逻辑。它们的存在打乱了人们正常的行进节奏，先是身

ity in urban space, it's particular communicative capacity. He devised a different method of remembering by installing mnemonic triggers rather than objects. Since the mid-1990s, he has been placing Stolperstein – stumbling stones – in sidewalks of Berlin in front the houses from which their Jewish residents were taken to concentration camps. Recently, the project expanded to include other victims of National Socialism: Roma, homosexual people and the disabled. The stones, made in size and shape of the commonly used pavement brick, bear names of the dead, the address at which they lived and were taken from, and the details of their deaths. Instead of securely placing names in the specifically designated areas, they are sunk into the road, walked over and made to be tripped on. The idea indirectly echoes the Jewish Haggadah, a guide to the Passover seder, which commemorates the Israelite's exodus from Egypt. The Haggadah asks the believers not to remember a historic event but to place themselves, once again, within the moment of divine intervention that the exodus itself is (not was!), and in a sense, to re-enact it[8]: every generation should consider themselves as if they themselves were rescued from Egypt. Part of daily practice in Berlin – walking – "stumbling stones" have a similar logic. They disrupt a flow of a normal routine bringing consciousness back to the body and then back to history. Feet, dogs and bikes trample over them, people bend over to read the inscription, and at this banal moment a crucial choice is made: does one curse in annoyance or pause with respect[9]? The scripture to the memorial book is what stumbling stones to the equestrian statue. Ecology activates identities by triggering different experiences. Memory is the awareness achieved through them. Seeing the city through the lease of an "archive" and an "ecology" also means paying attention to the ways in which memories are mass-mediated. There has been an ever-increasing importance of digital media platforms and online

Azeville Battery房屋，法国，1941年
Azeville Battery, France, 1941

饿鬼节，Selegie路，新加坡
Hungry Ghost Festival, Selegie road, Singapore

体意识到它们的存在，然后人们会记起那段历史。无数只脚、狗和自行车从其上踏过或碾过，人们俯身去读刻在这些石头上的文字，就在这最平凡普通的一刻，人们却要做出一个重要选择：是恼怒地咒骂还是出于尊重在这些逝者的名字前稍作停留[9]？刻在这些石头上的文字如同介绍雕像的纪念册，而铺在路面上的绊脚石则如同李将军骑马雕像一样。生态学通过触发不同的体验来激活认同。记忆则是通过认同实现的意识。

通过"城市档案"和"城市生态"来观察一座城市，也意味着需要注意承载记忆的大众媒体和方式。如今，无论是为了使人们记住什么，还是为了让城市成为一座值得纪念的城市，数字媒体平台和在线社交网络的重要性日益凸显。以手机智能应用程序为例，这些应用可以将历史上旧镜头投射到实际存在的城市空间。Instagram也起到同样的作用（译者注：Instagram照片墙是一款运行在移动端上的社交应用程序）。现如今，使游客拥有ICT（信息和通信技术）介导的博物馆和文化遗产地参观经历已经成为该行业的一部分。交互式的、扩展的体验功能大大增强了博物馆传统的"保存–展览"参观体验。不过，最近，整座城市俨然已经变成了一个数字媒体空间，使用应用程序，可以使用户置身于不同的时空当中。Urban Archive（城市档案馆）是为纽约市开发的一款应用程序，它在多维数字平台上开凿了一个可窥见该市历史的孔隙。当用户走过一栋历史悠久的建筑物时（即使这栋建筑已不复存在），其先进而复杂的界面就会推送通知。它会给人某些特定社区才有的徒步旅行体验，而这些社区的故事以前都是不为人所知的。人们也可以使用一个流行的并排照片

social networks for both, cultivating memories and making cities memorable as a whole. Take examples of mobile applications that project historical footage onto actually existing urban spaces; or the Instagram. ICT mediated museum and heritage-site experiences have been part of the business for a while now. They augmented the traditional "preserve-and-display" approach with interactive, extended experience feature. Recently, however, inroads were made into presenting an entire city as a media-space, and by doing so, transposing the user to different chronotopes (places-in-time) with the use of the app. Urban Archive, an app developed for New York City, creates apertures into the city's history in a multi-dimensional digital platform. Its sophisticated interface sends push notifications when a user walks past a historic building, including a no-longer existing one. It curates walking tours designed by specific urban communities whose stories were previously untold. It allows for a popular side-by-side photo generator to produce images which can be shared on popular social media. The emergence of such technologies not only enriches the perception of a city – it allows seeing what is no longer there, but also gives agency to individual urban explorers and enthusiasts, as much as communities and volunteer associations.

The ICT's another effect is on the individual memory of the user who takes snapshots, uploads his or her favourite images on the social media, the blogosphere or sends them to public image repositories. Together, these actions contribute to building a city's external image, its branding and trending. How much we can learn from the Instagram about a city remains a widely debated issue, of course. For instance, it is quite possible that proliferation of selfies and snapshots of specific, high-profile places, such as the Museumplein in Amsterdam, popularises so much, making it so iconic, that the complex and layered history of the location – for example, its connection to colonial

绊脚石，冈特·德姆尼希设计，捷克共和国科林
Stolpersteine by Gunter Demnig, Kolin, Czech Republic

生成器来生成可在社交媒体上共享的图像。这种技术的出现不仅丰富了人们对一座城市的感知——因为它能够使人们看到不复存在的东西，还为不同的城市探险家和爱好者提供了一个类似社区和志愿者协会的共享机构。

ICT的另一个影响体现在这种技术使用者的个人城市记忆方面，使用者拍摄快照，将最喜欢的图像上传到社交媒体、博客圈，或将图像发送到公共图像存储库。这些行为共同构建了一个城市的外部形象，打造了城市品牌和城市潮流。当然，我们能从Instagram上面获取多少关于一座城市的信息，这还有待进一步广泛讨论。举例来说，像阿姆斯特丹博物馆广场这样引人注目、知名度较高的场所的自拍和快照在社交媒体出现得越来越多，因此非常有可能越来越受人们的关注，成为一个标志，而使人们忽略了这类场所本身体现出的历史厚重感。例如，世界博览会让人们知道阿姆斯特丹博物馆广场与殖民贸易休戚相关。而现在，这类场所正逐渐沦为浮华之地，而非一个可以唤起人们关于历史记忆的场所。

这一切对城市的纪念传统（城市里的纪念碑和纪念堂）有什么借鉴意义？首先，我们需要打造一些灵活多变、能够适应不断变化的纪念场所。福特基金会主席、纽约城市艺术纪念碑和地标委员会联合主席达伦·沃克曾说道："我们民族有太多关于我们自身身份的记述，这反映了谁掌握权力、谁拥有特权。"[1]随着争取社会正义的斗争继续进行，纪念场所需要能够反映其动态和不断变化的权力格局，能够反映那些蕴藏在史书中现在变得明晰的道德态度，能够反映社会记忆的选择性。其次，纪念性建筑应该不仅仅被看作是用来存储记忆的时间胶囊，也应该被看作是公众的"刺激物"。当被问及像夏洛茨维尔那样的纪念碑或纪念雕像应该如何处理的时候，一

trade via the World Fair – escapes interest of the public. The place is being reduced to a glitz, not a lived and alive memory.

What are the implications of all this for commemorative traditions in a city, its monuments and memorials? First, such places need to be created with ongoing flows of change in the society in mind. They need to be flexible. In the words of Darren Walker, the president of the Ford Foundation who co-chaired the Commission on City Art, Monuments and Markers in New York, "So much about our narratives of who we are as a people is a reflection of who has power, who has privilege,"[1] and as the struggle for social justice continues, memorial sites need to be capable of reflecting its dynamics and the changing landscape of power; the moral attitudes implicit in the writing of history as those become explicit, and the very selectivity of social memory. Second, memorials should be seen not only as time capsules but also as public "irritants". When asked what should be done with the monuments like those in Charlottesville, an elderly woman said that they should stay put, so that her grandchildren could know what happened here. Their presence can ignite, rather than merely reflect, a public dispute, disagreement or conflict. Naturally, this may touch on the memorial's existence as a whole or its specific aesthetic qualities; some will become unwanted and removed, others altered. More important than a monument's preservation is a societal dialogue about history associated with it that the monument can ignite. "Flexible monument" might become a material expression of revisions of history natural to any society. (Of course, in some cases there is a sheer collective joy that comes from toppling of public statues, and that cannot be overlooked. Such acts in themselves are crucial

博物馆广场的荷兰国立博物馆,荷兰阿姆斯特丹
Rijksmuseum at the Museumplein, Amsterdam, the Netherlands

位老年妇女说它们应该被保留下来,这样她的子孙们才能知道这里曾发生过什么。这些纪念性建筑的存在不仅仅能反映历史,还会引发公众争论、分歧或冲突。自然而然地,这可能关系到纪念性建筑的存在,或关系到纪念性建筑特定的审美品质;有些会变得多余,不再被需要而被移除,有些则需要做出调整和改变。相较于纪念性建筑的保存,更为重要的是它本身可以引发一场与其相关的关于历史的社会对话。"灵活的纪念碑"可能成为任何社会修正历史的一种实物表达。(当然,在某些情况下,推倒公众雕像会带来纯粹的集体喜悦,这是不容忽视的。这些行为本身就是创造历史的关键时刻。)

随着城市里一些地方被认可为记忆生态,一切可以更有效地进行。每件事都与某件事有关,有其自身含义。富有纪念意义的地点钩织成相互联系的网络。为什么不使用新加坡人的方法来处理他们被抹去的家庭历史,将大屠杀的夜间地形投射到德国和东欧的城镇地区?或者是通过ICT技术将民权运动的数字化图像与南部邦联的雕像并列呈现?嗅觉和听觉刺激都可以被广泛地用来带动人们的感受体验,唤起记忆,并且通过这样的方式来让人们想起那些早已被遗忘的事物。相较于遗忘,人们更感兴趣的显然是记忆,现在也许是时候改变这种不平衡,使"遗忘"以自己的方式存在于城市景观中。毕竟,"旧的方式"在城市的物质文化中有自己的联系,而对于一些真正要被废弃的传统来说,消除可能是必需的。

moments in the making of history.)
With the recognition of urban places to be ecologies of memory, this can be done more effectively. Everything relates to something and leads somewhere. Mnemonic sites are connective fabrics. Why not use the approach of Singaporeans to their erased family histories and project night-time topographies of the Holocaust onto neighbourhoods of German and East European towns; or the ICT capabilities to digitally juxtapose the imagery of the Civil Rights movement over the Confederate statuary? Olfactory and sonic stimuli can be summoned more extensively to bring out experiences, evoke memories, and by doing so invoke presences that have been long forgotten. And while memory is often considered a more interesting question than forgetting, it may be the time that this imbalance is corrected and "forgetting" finds its own form in the cityscape. After all, the "old ways" have their own wiring in material culture of the city, and for some traditions to be truly dispelled an erasure might be imperative.

1. Jelani Cobb. 'New York's Controversial Monuments Will Remain, But Their Meaning Will Be More Complicated'. The New Yorker. January 12, 2018.
2. Sennett, R. *The Conscience of the Eye. Design and the Social Life in the City*. W.W. Norton Company, Inc. 1992. p.39.
3. David Harvey. 'Monument and Myth', Annals of the Association of American Geographers Vol 69, No 3 (Sep., 1979). pp.362-381.
4. Ladd, B. *The Ghosts of Berlin*. Chicago and London: The University of Chicago Press. 1997.
5. Sezneva, O. 'Your Place, My Memory'. Unpublished manuscript.
6. Alois Riegl. 'The Modern Cult of Monuments: Its Character and Its Origin'. 1903. Trans. Kurt W. Forster and Diane Ghirardo.
7. Comaroff, J. 'Ghostly topographies: landscape and biopower in modern Singapore', Cultural Geography, Vol. 14, Issue 1, January 2007, pp.56-73.
8. Gil Eyal, 'History, memory, identity. Two Forms of the Will to Memory'. History & Memory, Vol 16, No. 1, 2004. pp.5-36.
9. 'Stumbling Upon Mini Memorials To Holocaust Victims'. NPR, Morning Edition. May 31, 2012.

记忆之家
House of Memory
baukuh

记忆之家是米兰市的一个档案馆,也是一个举办展览和会议的地方。作为五大文化协会(其中包括意大利纳粹灭绝集中营政治驱逐者全国协会等)的总部,其目的是保存意大利对自由与民主的征服的记忆。当代的米兰没有一个稳定的完全可共享的记忆,一个可以雕刻在石头上而不会受到诘难和质询的记忆。建筑师事务所baukuh表示,与其把记忆之家当作一个表达共享记忆的场所,他们更愿意把它看作是一个媒介,在此,人们可以讨论共存于这个城市集体记忆之中的各种元素。记忆之家试图为各种不同的记忆提供一个庇护的场所。这些记忆不仅与当代社会交织在一起,也深植于每个人的内心之中。

记忆之家外墙完全由赤土瓷砖建造。这些瓷砖构成了一幅幅描绘米兰近代史的巨型画像。这座新建筑的外壳被人们视作一个多联画屏,展现了米兰战后记忆中的重要人物与事件,也为展示被定义为过去的共同时刻提供了一个机会,这一时刻不仅与记忆之家中的五大文化协会有关,也与所有米兰市民有关。文化协会与委员会制作了一个网站,首先请市民们推荐建筑立面所采用的图像,然后请市民们从中选出最终采用的图像。图像确定之后,由艺术家们将其描绘在记忆之家的建筑立面上。

记忆之家是一座非常简单的建筑，一座长35m、宽20m、高17.5m 的长方体建筑。该建筑分为三部分，通过完全开放的一层连为一体。这座长方体建筑的两端分别是两块较窄的空间，设有档案室（靠南面）、洗手间和技术设备室（靠北面）以及垂直交通空间。一层的开放空间被由两根八角形柱子划分为三部分。该区域的三分之一采用通高设计，有一部螺旋形楼梯。其余部分为展厅和办公室，分布在三个楼层。该建筑的内部组织——巨大的黄颜色楼梯镶嵌在分布在三个楼层的办公室和展厅以及分布在五个楼层的档案室之间——使得建筑内部空间非常宽敞。局促狭窄的档案室空间和巨大的楼梯体量形成鲜明对照，这样的反差使得办公室和展厅空间显得格外宽敞，让访客感到非常宽敞明亮。黄颜色楼梯不仅是该建筑主要的交通通道，也是使访客和馆藏之间建立联系的重要设施。鉴于档案的珍贵性，访客不能直接接触文件，市民和馆藏之间的关系是通过楼梯旋转产生的运动建立起来的。访客不断地与馆藏靠近而又远离，从而经历了复杂的视觉变化，一会儿看到馆藏文件，一会儿看到外面的公园。

光线以多种方式进入建筑内部。办公室区域有几个巨大窗户，集中分布在建筑外立面的相应部分。建筑开放的空间布局以及用玻璃隔断细分内部空间的方法都保证了光线的充足性。入口大厅、巨大的楼梯以及一层空间均可通过几个大的洞口接收光线；光线照进半明半暗的建筑空间内，营造出了一种宁静庄严的氛围。

记忆之家坐落于米兰伊索拉地区的波尔塔努欧瓦。该建筑外观粗糙，所使用的建筑材料低调、冷艳，体现了当地伦巴第传统，与伊索拉地区的历史建立了深刻而紧密的联系。由于旧工厂的消失以及近期在伊索拉地区进行的旧区改造，这些协会将这个使该地区与过去保持联系的唯一元素保留了下来。从大的方面来说，记忆之家旨在建立与整座城市的历史的联系；而从地方层面来说，该建筑需要做的是与其邻近地区历史和身份建立一种更紧密的联系。

The House of Memory is an archive as well as an exhibition and conference space in Milan. It is the headquarter of five cultural associations including the National Association of Italian political Deportees in Nazi extermination camps (A.N.E.D.), etc., whose aim is to preserve the memory of the conquest of freedom and democracy in Italy. Contemporary Milan does not possess a stable, entirely shared memory, ready to be carved in stone without further interrogation. Rather than considering the House of Memory as an expression of shared memory, the architects, baukuh, preferred considering it as a tool for discussing the different elements that coexist within the collective memory of the city.
The House of Memory tries to provide a shelter for the various and varied memories that are woven not only into contemporary society, but also in the minds of individuals. The House of Memory is entirely covered in terracotta tiles that compose large images depicting the recent history of Milan. The shell of the new building, understood as a polyptych to represent the fundamental people and events of Milan's post-war memory, provides an opportunity for a

collective moment in the definition of the past, a moment that involved not only the associations comprising the House of Memory, but all citizens. The associations and the committee produce a website that allows citizens first to propose images, and then later to select images from the ones proposed. Once the images are defined, the artists depict them on the surface of the House of Memory.

The House of Memory is a very simple building: it is a box with a rectangular base of 20m by 35m and 17.5m high. The building is divided into three parts that are connected to one another by an entirely open ground floor. Two thin layers along the building's shorter ends house the archive (south), the restrooms and technical installations (north), and the vertical circulation. The open space on the ground floor is subdivided into three parts by two octagonal columns. One third of this area reaches the building's full height and includes a spiral staircase. The rest is occupied by exhibition spaces and offices disposed on three levels. This internal organization – with the enormous, yellow staircase inserted between the three levels of offices and exhibition spaces and the five levels of the archive – introduces a greater scale into the building. The contrast between the tight levels of the archive and the colossal dimension the staircase allows the office and exhibition spaces to acquire spaciousness; visitors perceive a vaster, more generous atmosphere.

The yellow staircase is not only the building's main distributive element, but also the device that establishes a relation among visitors and the collection. Given that the preciousness of the archive does not allow visitors to directly access to the documents, the relationship between the citizens and the collection is established through the rotating movement created by the staircase. Visitors repeatedly come closer to and then move away from the collection, thereby experiencing a complex sequence of views of the documents and of the park outside.

Light enters the building in many fashions. The offices have large windows concentrated in the corresponding parts of the facade. The building's open-space configuration and internal subdivision with glass emphasize the abundance of light. The entrance hall, the large staircase and the ground floor receive light from a handful of very large openings; grazing light invades the semi-darkness and generates a calm and solemn space.

The House of Memory is located in the Porta Nuova neighbourhood in the Isola Quarter of Milan. The building's rough appearance and the proud sobriety of its humble construction materials, which reflect local Lombard tradition, establish a deep and precise connection with the history of the Isola Quarter. Due to the disappearance of the old factories and the gentrification in the Isola Quarter's recent transformation, the associations have remained the only element that maintains a relationship with the area's past. If, on a large scale, the House of Memory aims at establishing a relationship with the entire city's history, then on a more local scale, the building needs to establish a precise relationship with the history and identity of its immediate neighbourhood.

记忆游戏——建立肖像程序
Memorial Game – setting-up the iconographic program

第一阶段：协会和市政当局指定科学委员会/第二阶段：科学委员会和市政当局建立图像档案馆
第三阶段：市民选择图像/第四阶段：艺术家（由科学委员会指定）制作立面图像
Phase 1. the associations and the Municipality appoint the Scientific Committee / Phase 2. the Scientific Committee and the Municipality set-up the images archive
Phase 3. the citizens choose the images / Phase 4. the artists (appointed by the Scientific Committee) make the images on the facades

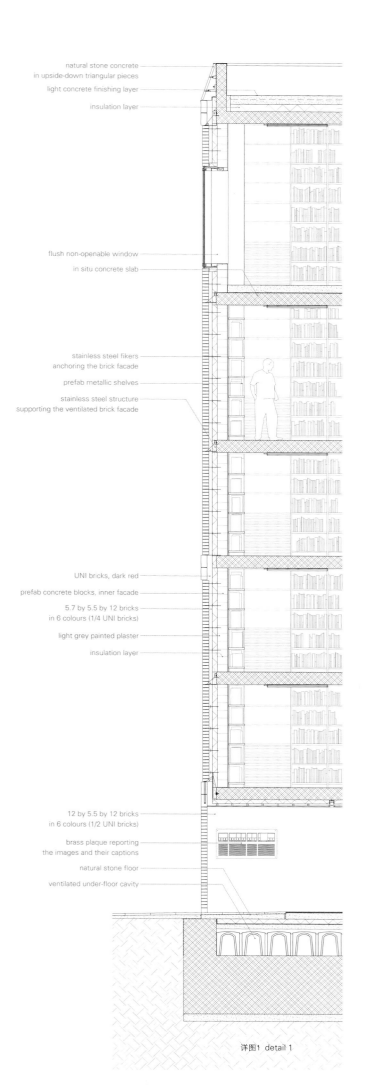

详图1 detail 1

-3.6m　　0.0m　　+3m　　+6m　　+7.5m　　+9m　　+13.3m　　+15.2m

- 520m² 档案室 archives 520mq
- 120m² 参考文献室 reference room 120mq
- 100m² 大厅 hall 100mq
- 185m² 展览区 exhibition area 185mq
- 185m² 多功能空间 multipurpose space 185mq
- 780m² 办公室 offices 780mq
- 65m² 酒吧 bar 65mq
- 80m² 浴室 bathroom 80mq
- 420m² 露台 terrace 420mq
- 50m² 储藏室 storage 50mq
- 200m² 无人技术空间 not habitable technical space 200mq
- 420m² 未来扩建空间 space for future expansion 420mq

公共空间的灵活性 Public Spaces Flexibility

一层 ground floor

自助餐演讲　　　慈善晚宴　　　　音乐会　　　　展览　　　　舞会
lecture with buffet　charity dinner　　concert　　exhibition　ballroom dancing

露台 terrace

 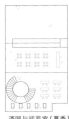  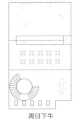 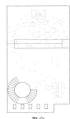

酒吧与阅览室（冬季）　酒吧与阅览室（夏季）　餐前鸡尾酒　周日下午　聚会
bar and reading　　　bar and reading　　　aperitif　sunday afternoon　party
room (winter)　　　room (summer)

办公室的灵活性 Offices Flexibility

三层和四层
2nd and 3rd floor

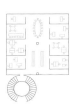

项目名称：House of Memory
地点：Italy, Milan
建筑师：baukuh architetti associati
结构工程师：Arup Italia srl
暖通空调：Deerns Italia spa
防火工程师：Gaeengineering srl
工料测量师：J&A consultants srl
施工公司：Edilda Edilizia Lombarda spa
客户：City of Milan
开发商：HINES ITALIA SGR SpA
公共空间面积：550m²
办公室面积：800m²
档案室面积：320m² (160,000 volumes)
设备空间面积：200m²
总面积：2,500m²
造价：EUR 3,600,000
竞赛时间：2011
详细施工设计时间：2013
施工时间：2013—2015
摄影师：©Stefano Graziani (courtesy of the architect) - p.14~15, p.16~17, p.18, p.20~21, p.22~23;
©Giulio Boem (courtesy of the architect) - p.19

1. 大厅	1. hall
2. 演讲厅/展厅	2. lectures/exhibition
3. 技术设备室	3. technical room
4. 卫生间	4. toilet
5. 档案室	5. archive
6. 办公室	6. office

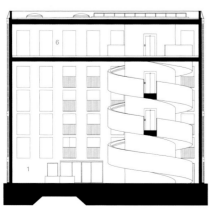

A-A' 剖面图 section A-A'

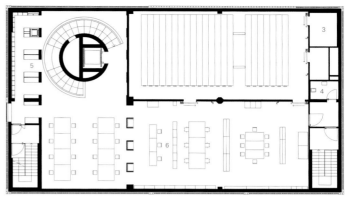

四层 third floor

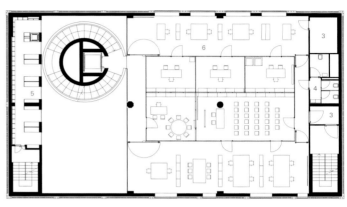

二层 first floor

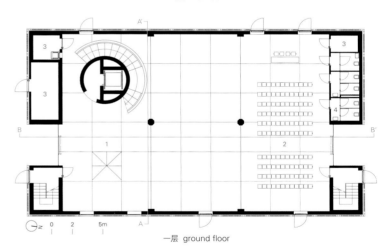

一层 ground floor

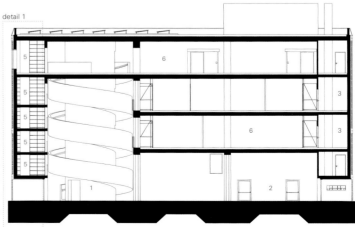

B-B' 剖面图 section B-B'

阿尔托大学主楼 Dipoli 大楼
Dipoli, Aalto University Main Building
ALA Architects

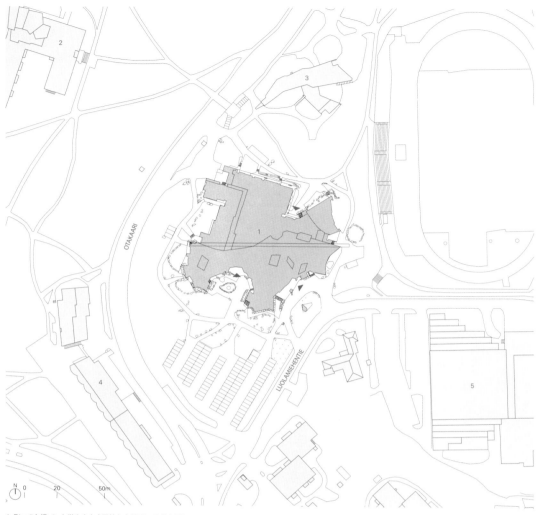

1. Dipoli大楼 2. 大学生中心（原赫尔辛基理工大学主楼）
3. Teknologföreningen楼——瑞典语学生的学生会大楼 4. 购物中心 5 Otahalli体育中心

1. Dipoli 2. undergraduate center (formerly the main building of Helsinki University of Technology)
3. Teknologföreningen - student union building of the Swedish speaking students 4. shopping center 5. Otahalli sports center

Dipoli大楼是一座列管建筑，具有一定的象征性和实验性，曾是赫尔辛基理工大学的学生会大楼，由Raili Pietilä和Reima Pietilä设计，于1961年赢得设计竞赛，于1966年建成完工。最初，Pietilä夫妇的设计方案在竞赛中获得了并列第二名，而后在为两个并列第二名举行的第二场竞赛中获胜。在改建之前长达20年的时间中，Dipoli一直作为会议中心使用。

经过一次彻底的改建，该建筑重新焕发了生机，成为阿尔托大学奥塔涅米校区的主教学楼，并于2017年秋季学期重新开放。新的校区将与原先的赫尔辛基理工大学校区合并为阿尔托大学的主校区——后者本身就是由赫尔辛基的三所大学合并而来的。设计团队的目标是在尊重原设计者设计理念的基础上，通过构建全新、开放、动态的使用体验，使Dipoli大楼重新焕发活力。Dipoli大楼已成为大学行政部门、学术团体、学生们和相关人员的集会场所，各方都参与了重新设计的过程，创造了一个未来可持续的、灵活的工作空间。除了容纳行政办公场所之外，大楼还将继续作为举办重要讲座和节日活动的场地，同时为学校的研究成果和设计作品提供展示平台。Dipoli大楼的餐厅、自助餐厅和酒吧也将向学生和工作人员开放。

作为阿尔托大学灵活的工作方式和可移动的工作环境的"实验室"，Dipoli大楼将为该大学行政部门的200位员工提供工作场所。

Dipoli, the listed iconic and experimental student union building of Helsinki University of Technology was designed by Raili and Reima Pietilä and completed in 1966 after the competition in 1961. The Pietiläs' entry was originally awarded shared 2nd prize and later selected as the winner of the second competition organized between the two 2nd prize winners. Prior to the renovation, Dipoli functioned as a conference center for a period of 20 years.

Through a complete renovation, the building has now

gained a new life as the main building of Otaniemi campus of Aalto University and reopened for fall semester 2017. The renovation was part of the larger campus reorganization project linked to the former Helsinki University of Technology campus becoming the main campus of Aalto University, born out of the merger of three Helsinki area universities. The design team's aim was to re-radicalize Dipoli by creating a fresh, open and dynamic user experience, not forgetting the original designers' vision. Dipoli has become a meeting place for the university administration, the academic community, the students, and other stakeholders. All of these parties have been activated in the spatial re-design process that turned the building into a sustainable, flexible workspace of the future. In addition to housing the administration, Dipoli also continues to function as the prime location for important lecture events and university festivities, as well as acts as a display platform for the university's research and design projects. Dipoli's restaurants, cafeterias and bars cater for both students and staff members.

Dipoli is Aalto University's test lab for flexible working methods and mobile work. Two hundred of the university's administrative employees will use the building as their base.

南立面 south elevation

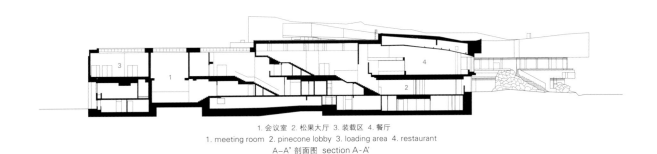

1. 会议室 2. 松果大厅 3. 装载区 4. 餐厅
1. meeting room 2. pinecone lobby 3. loading area 4. restaurant
A-A' 剖面图 section A-A'

1. 活动大厅 2. 主厅门厅 3. 大厅
1. events lobby 2. lobby for main hall 3. hall
B-B' 剖面图 section B-B'

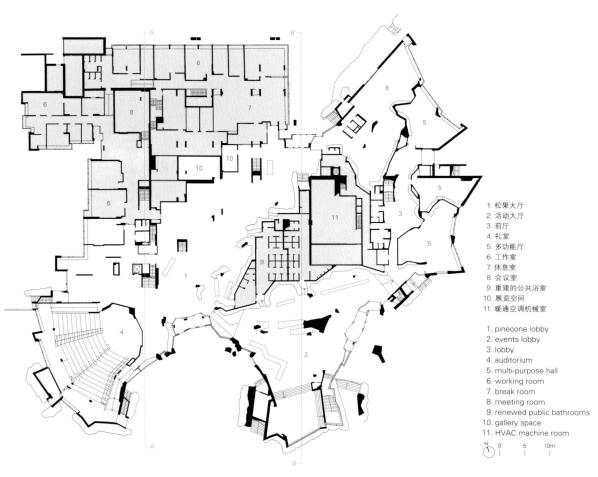

1. 松果大厅
2. 活动大厅
3. 前厅
4. 礼堂
5. 多功能厅
6. 工作室
7. 休息室
8. 会议室
9. 重建的公共浴室
10. 展览空间
11. 暖通空调机械室

1. pinecone lobby
2. events lobby
3. lobby
4. auditorium
5. multi-purpose hall
6. working room
7. break room
8. meeting room
9. renewed public bathrooms
10. gallery space
11. HVAC machine room

二层 first floor

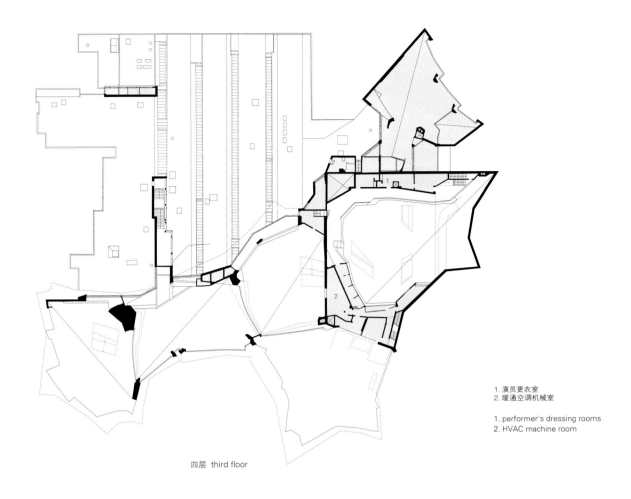

1. 演员更衣室
2. 暖通空调机械室

1. performer's dressing rooms
2. HVAC machine room

四层 third floor

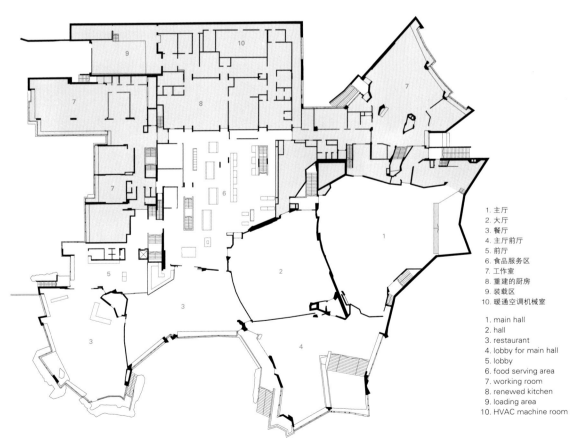

1. 主厅
2. 大厅
3. 餐厅
4. 主厅前厅
5. 前厅
6. 食品服务区
7. 工作室
8. 重建的厨房
9. 装载区
10. 暖通空调机械室

1. main hall
2. hall
3. restaurant
4. lobby for main hall
5. lobby
6. food serving area
7. working room
8. renewed kitchen
9. loading area
10. HVAC machine room

三层 second floor

项目名称：Dipoli, Aalto University Main Building / 地点：Otakaari 24, Espoo, Finland / 建筑师：ALA Architects
项目团队：Juho Grönholm, Antti Nousjoki, Janne Teräsvirta, Samuli Woolston, Toni Laurila, Pekka Sivula, Simo Nuojua, Lotta Kindberg, Tiina Liisa Juuti, Marlène Oberli-Räihä, Mirja SIllanpää and Sari Vesanen / 合作者：Office space concept – Workspace; Service design – Creadesign; Conservation – Kristo Vesikansa; Building services engineering – Ramboll Finland; Fire safety – Palotekninen insinööritoimisto Markku Kauriala; Structural design – Vahanen Group; Interior design – Tuuli; Sotamaa; Acoustics design – Helimaki Acoustics; Kitchen design – Suurkeittiö-Insinööritoimisto Rita Pulli; Main contractor – NCC Building

供应商：Fixed furnishings - Kiimingin kaluste; Door restoration - Suur-Helsingin Kirvestyö; Window restoration - Vanalinna Ehitus; Restoration of auditorium chairs - Idea-Puu; Restoration of lighting fixtures - Artisan Rinaldo; Lowered ceiling in student cafeteria - OUTrading; Stage velours - Melodrama / 客户：Aalto University Properties / 用途：offices, restaurants, bars, auditorium, conference rooms, meeting rooms, open space, faculty club, etc. / 用地面积：18,000m² / 建筑面积：11,400m² (usable floor area after renovation) / 总楼面面积：12,400m² (gross floor area after renovation) / 竣工时间：2017 / 摄影师：©Tuomas Uusheimo (courtesy of the architect)

海法大学 Younes & Soraya Nazarian 图书馆
University of Haifa,
Younes & Soraya Nazarian Library

A. Lerman Architects

在位于卡梅尔山脊顶部的新海法大学校园能够俯瞰海法这个港口城市的全部，巴西建筑师奥斯卡·尼迈耶在对这个项目进行设计时说道："这个地方美得让人窒息，显然需要一个紧凑的设计方案，简单而又宏伟。"他的设计彻底改变了该地原来的自然环境，在一个巨大的矩形平面上创造了一个人工景观，为卡梅尔山脊创造了一个新的人造轮廓，庞大而宏伟，成为海法大都会的标志，人们既爱又恨。

尼迈耶的设计方案与传统意义上的"校园"建筑类型背道而驰。他的设计是将所有分散的布局聚集到一个单一的结构当中，将校园的所有功能设施纳入一个巨大的屋顶之下。公共通道与大厅的位置可以划分不同的功能设施，同时也是校园师生的集会场所。这座独特的大学有来自以色列北部各地的学生，他们属于不同的种族（犹太人占60%，阿拉伯人占40%）。正如Zvi Elhyani在其《奥斯卡·尼迈耶和1960年后以色列投机城市主义的开端》一文中所提到的那样，尼迈耶关于"最大程度互动的跨学科多元学术"的愿景得以成功实施。这个巨大的基座和塔楼包含了图书馆和其他设施，周围布置有一系列的走廊与公共空间。

图书馆大楼建成开放已有45年，它见证了诸多变化。图书馆大楼长270m，宽70m，非常开阔，但随着学生与教职工数量的不断增加，图书馆大楼也在不断变化。结果，原来流畅的开放空间变得拥挤不堪。没有经过合理规划布局的设施影响了采光，根本没有方向感而言。

设计任务书要求，所提出的设计方案要解决空间拥挤、采光不好、空气不流通等问题。另外，任务书中要求设计一处供图书馆工作人员使用的设施。任务书还明确指出图书馆的"中央核心"功能——既是一个多元化的信息空间，又是师生见面交流的地方。

为了实现设计多样性，而不是强制实现对话联系，建筑师分析了该建筑目前所面临的两个问题：原设计的僵硬呆板和不合时宜以及目前毫无设计感的迷宫般的混乱。

为了恢复原设计开阔的布局，使阳光从东面射入室内，建筑师将这座巨型建筑内部后加的隔墙尽数拆除，在主要位置设计了一个通往图书馆和这个"中央核心"的新入口。紧邻该建筑的东立面设计有一个开放式庭院，它主要起到通风换气的作用。这是个狭长的空间，110m长，仅8m宽，一侧是原有建筑，另一侧是新建的图书馆的翼楼。这座新建的翼楼位于地下，上面是停车场，只有一个立面面向庭院。实际上，从当前图书馆的任何地方都无法看到它。与原本没做任何改变、看似由无穷无尽的直线构成的立面相反，新的立面扭转弯曲，创造出一些独特的供人们休息独处的小型空间。因此，庭院空间成为第三个空间，一个自主空间，吸引着人们加入这个位于塔楼垂直景观下面、两座不同建筑之间的空间对话之中。

"阿萨夫·勒门设计的图书馆与尼迈耶的宏伟设计截然相反。然而，矛盾的是，这也是一个重要的充满智慧的建筑作品，成功地继承了尼迈耶激进的现代主义，不像多年来堆建在校园里的许多其他后现代主义建筑，它们完全与原来现代主义建筑风格背道而驰。"以色列建筑档案馆馆长Zvi Elhyani博士如是说。

Approaching the design of the new Haifa university campus in a site overlooking the port town from the top of the Carmel ridge, Oscar Niemeyer, a Brazilian architect, stated: "The overwhelming beauty of the site clearly calls for a compact solution, simple but monumental." His scheme cleared away the site from its natural context and created an artificial landscape in the shape of a giant rectangular plain. He created a new artificial and monolithic contour to the Carmel ridge that would become the much loved/hated symbol of the Haifa metropolis.

Niemeyer's design was in direct opposition to the conventional architectural typology of the "campus". His proposal was a condensation of its dispersed layout into a single structure designed to accommodate all programs under a single giant roof. Public passages and halls were laid out in order to differentiate programs from one another and act as meeting places for the campus dwellers. In this unique university, collecting students from all over the north of the Israel and creating a mixed racial (60% Jew and 40% Arab) body of students, Niemeyer's vision of an "inter-disciplinary plural academism of maximum interaction" was implemented successfully, as Zvi Elhyani mentioned in *Oscar Niemeyer and the Outset of Speculative Urbanism in Israel after 1960*. The giant base and tower contained the library and other facilities arranged around a series of hallways and communal spaces.

In the 45 years that have passed since the library's opening the building witnessed many changes. Its 270m long by 70m wide, open plan, has been repeatedly changed as the building served an ever-increasing number of students and

1 原有建筑
1. existing

2 清空
2. emptying

3 重新填充
3. re-filling

4 新的空间关系
4. new spatial relationships

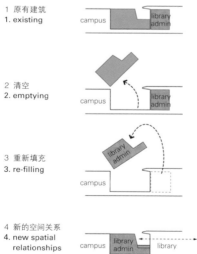

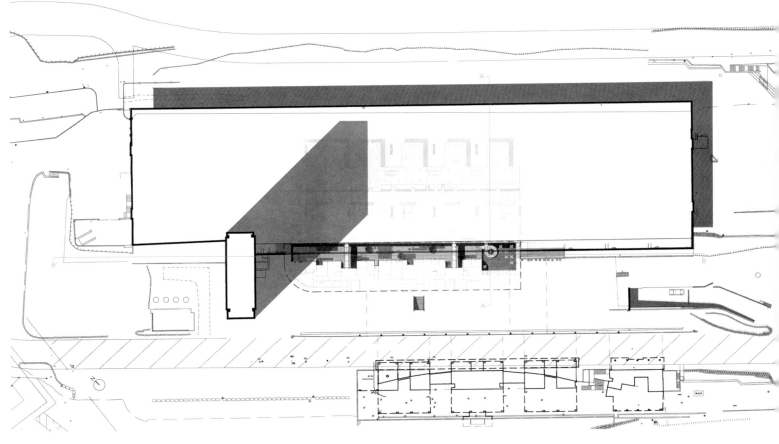

staff. As a result, the fluent open spaces had been compressed by ill planned partitions blocking off the sunlight and rendering the orientation impossible.

The competition brief demanded a solution for the problem of crowdedness and lack of sunlight and air. It included a detailed program for an additional building that would accommodate the library's staff. It also specified the library's need for a "central core" as a diverse information space and a place for encounter between students and staff.

Both "ends" of the building, the rigid romantics of the original plan, as well as the chaotic maze of its current "un-plan", were analyzed in order to articulate an intervention that was aiming at diversity, rather than forcing dialogue.

Inside the mega-structure the added partitions were cleared off in order to re-instate the open plan and allow the sunlight in from the east. It allowed for the introduction of a new main entrance to the library and to the central core to occupy the primary position. Adjacent to the eastern facade, a technical ventilation courtyard was re-claimed as an open patio. This dramatic 110 meters long space is 8 meters wide only, bordering on one side to the existing building and on the other to the new library wing that was dug underneath the elevated parking lot. The new subterranean wing has only one facade that is facing the patio and it is in fact invisible from anywhere asides the existing library. In contrast to the seemingly endless straight line of the original facade that is left untouched, the new facade twists and bends so to create particular small scale hideaways. Thus the space of the patio becomes a third and autonomous category of space that invites visitors to participate in the spatial dialogue between the two opposing sides underneath the vertical landscape of the tower.

"The Library designed by Asaf Lerman is an architectural statement that is diametrically opposed to Niemeyer's monumental plan. Paradoxically, however, it is also a critical and intelligent architectural tribute that successfully takes on Niemeyer's radical modernism, unlike the dozens of post-modernist structures that were heaped upon the campus through the years, irreversibly undermining its original qualities," said Dr. Zvi Elhyani, Founding Director at Israel Architecture Archive.

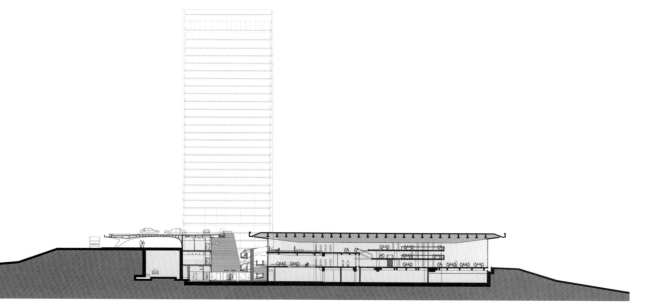

A-A'剖面图 section A-A'

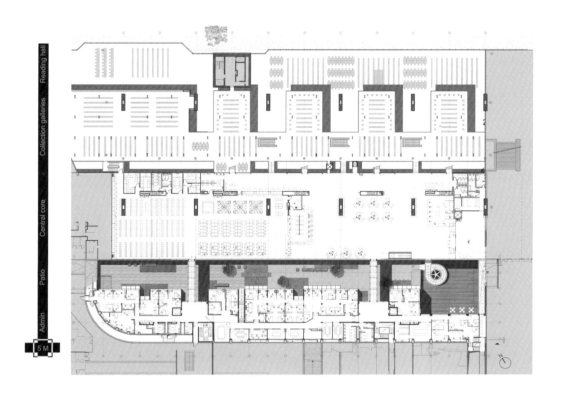

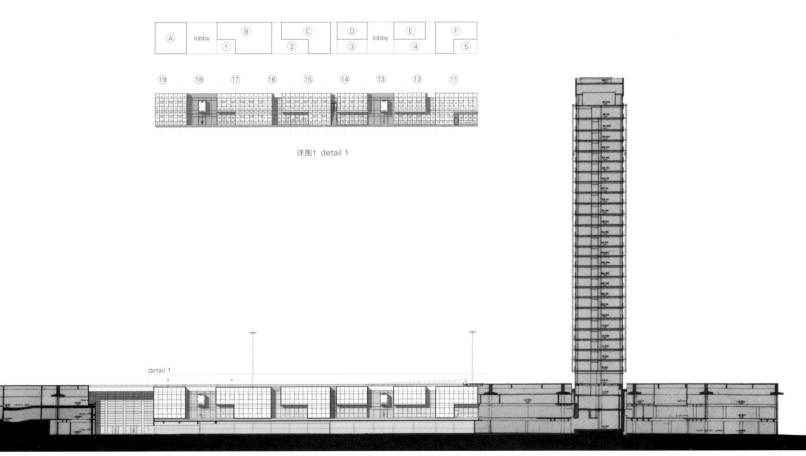

详图1 detail 1

B-B'剖面图 section B-B'

项目名称：University of Haifa Younes & Soraya Nazarian Library / 地点：Haifa, Israel / 建筑师：A. Lerman Architects / 项目团队：Asaf Lerman, Lev Konikov, Nimrod Schenkelbach / 项目管理：Yakov Shaked - Lavid Engineering Ltd. / 结构工程师：Muller - Shnabel - Tzahar
机械工程师： AM Insertional - Avi Menashe / 电气工程师：TOPAZ group / 照明工程师：TOPAZ group
景观建筑师：A. Lerman Architects / 承包商：Abu-Ayash / 客户：University of Haifa / 用途：library
用地面积：existing bldg - 4,000m²; new bldg - 1,800m² / 建筑面积：existing bldg - 4,000m²; new bldg - 5,000m² / 建筑覆盖率：85%
结构：concrete / 室外饰面：concrete + glass (printed) / 材料：concrete, wood, steel / 造价：$17,000,000
设计时间：2007—2011 / 施工时间：2009—2013 / 竣工时间：2013 / 摄影师：©Amit Giron (courtesy of the architect)

少女塔的修复
Restoration of the Maiden Tower

De Smet Vermeulen Architecten + Studio Roma

少女塔是一座建于14世纪的城堡的主塔，但是部分已经坍塌。由于缺乏相关历史文献，所使用的建筑材料（当地的一种铁石）已经没有了，要重建塔楼几乎是不可能的。修复塔楼的目的是加固废墟，防止其再进一步坍塌，同时也为人们提供一个观景平台，以欣赏周边的风景。塔楼的修复基于结构逻辑：几米厚但部分倒塌的墙壁需要加一层顶盖，外部呈圆柱体、内部为八角形的塔身需要进行修复以加固整个结构。

一条有顶的小路通往进入少女塔必经的一座小桥；顶盖结构可以保护游客免遭坠落碎石砸伤。进入塔内要经过19世纪修建的入口，这个入口至今仍然稳固。一段长长的金属楼梯通向上层，在那里有另外一段楼梯填补了外墙的缺口，楼梯平台则插入了加固后的塔身外墙（其中注入了合成树脂）与里面的空隙。在塔楼里面，混凝土墙墩取代了已经坍塌的交叉拱顶；而在外部，砖壳体封闭了倒塌的圆柱体塔身。砖块大小的洞口为塔楼内新建的楼梯提供光线，这样就无须在外墙上开新的窗户，而且也无法考证历史上少女塔是否曾有这样的窗户。

通往屋顶的最后一部楼梯是旋转楼梯，与在原墙体上发现的一节楼梯连为一体。整个钢结构建于坍塌的墙壁和原几米厚的墙壁之上，呈现出与原塔身一样的圆形轮廓，形成一个与罗马坦比哀多礼拜堂类似的结构，其上层可以用作瞭望台。

The Maiden Tower is a partially collapsed, fourteenth-century donjon. Reconstruction is virtually impossible because the historical documentation is lacking and the building material, a local ironstone, is no longer available. The aim of the restoration was to reinforce the ruin, prevent its further deterioration and render the tower accessible as a viewing platform for the surrounding landscape. The restoration

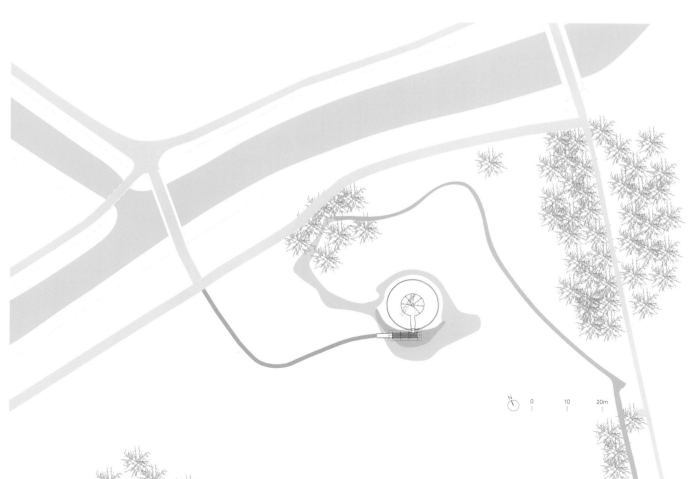

was based on structural logic: the meters-thick, crumbling walls needed to be capped and the cylindrical, internally octagonal geometry needed to be repaired to stabilize the structure.

A covered path leads to an access bridge; the cover protects visitors from falling debris. Access is via the stable nineteenth-century entrance. A long metal staircase leads to the upper floor where another staircase fills the breach in the outer wall. The stair landings plug the gap in the reinforcement rings of injected synthetic resin in the donjon walls. On the inside, concrete piers replace the collapsed cross vaults, and on the outside, a brick shell encloses the cylinder. Brick-sized openings illuminate the new stairs without the need to perforate the elevations with new and historically unvalidated windows.

The final stairway to the roof is a spiral stair, built into a stair shaft discovered in the wall. The steel structure that raises the meters-thick wall coping above the crumbling wall and that in silhouette restores the circle, generates a tempietto whose upper level serves as an observation post.

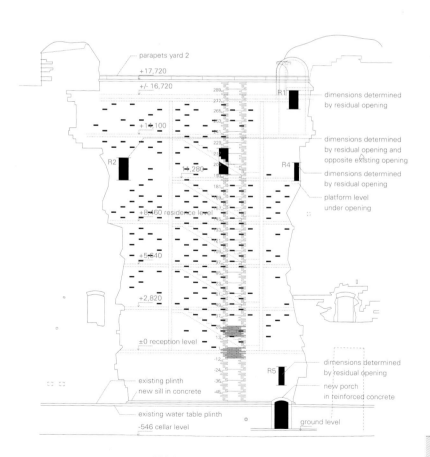

西南立面 south-west elevation

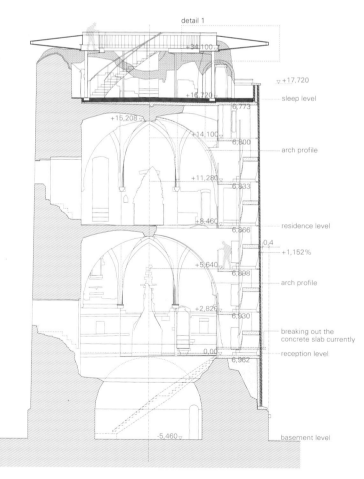

A-A'剖面图 section A-A'

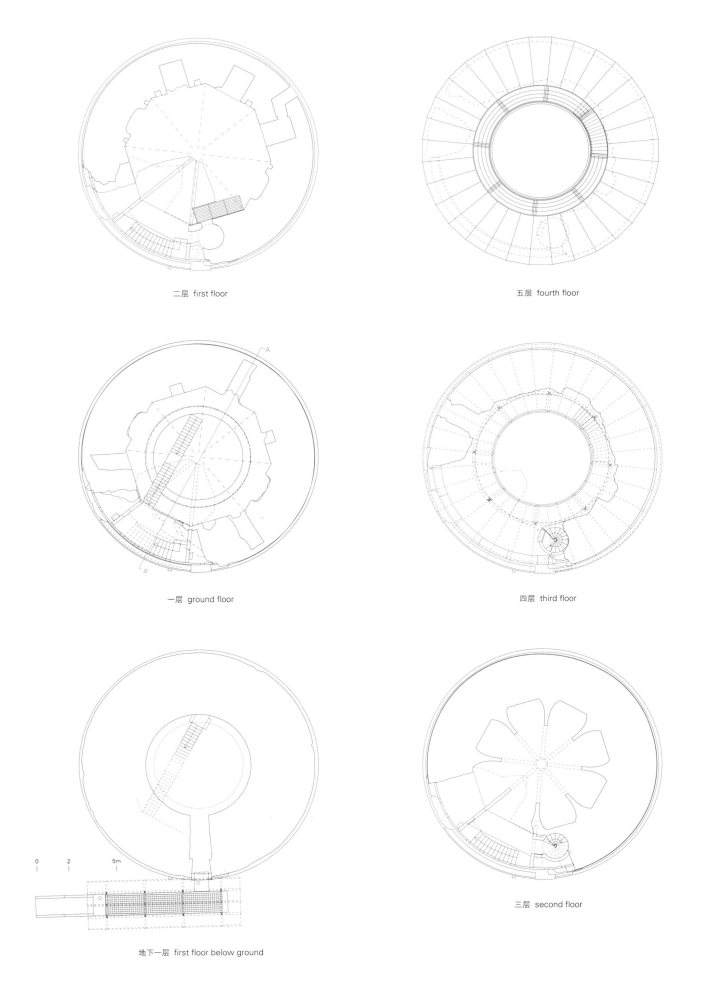

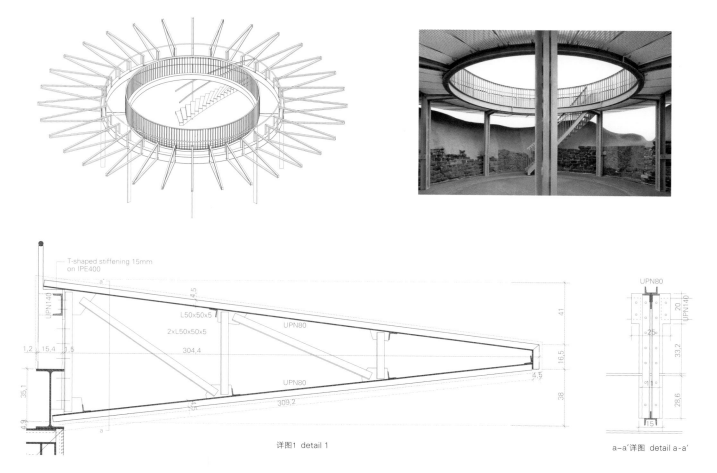

详图1 detail 1

a-a'详图 detail a-a'

项目名称：Restoration of the Maiden Tower
地点：Ernest Claesstraat 3271 Zichem, Belgium
建筑师：De Smet Vermeulen Architecten + Studio Roma
结构工程师：Norbert Provoost-Probam
设备工程师：Linda Van Dijck-Tecon
总承包商：Denys-Building; Monument Vandekerckhove with Altitempi
客户：Vlaamse Overheid, Agentschap onroerend erfgoed
用地面积：13,400m²
建筑面积：744m²
总楼面面积：2,976m²
竣工时间：2016
摄影师：©Filip Dujardin (courtesy of the architect)

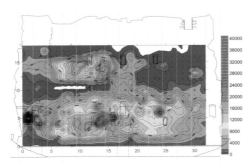

参考图表 reference sheet

安吉·雷迪博士纪念园
Dr. Anji Reddy Memorial

Mindspace

卡拉姆·安吉·雷迪博士是一名科学家、企业家,也是一名慈善家,是一个颇具传奇色彩的人物。他对于发明新药物有着极大的热情,他努力使药物让病患消费得起,在控制药价方面是一位先驱者。安吉·雷迪博士出生于印度的Tadepalli,在印度一家国营制药公司工作了六年,随后在1984年,他建立了雷迪博士实验室。

他一生致力于通过制造人们负担得起的药物来减缓病人的痛苦,尤其关注医药研究和临床应用。他富有同情心,积极尝试重新定位慈善,重新思考如何发掘弱势群体的潜力,堪称楷模。

雷迪博士纪念园带我们思考和领略雷迪博士一生的不同侧面,向人们传达出如下独特而大胆的信息:无论我们的出身有多卑微,我们

的命运都掌握在自己的手中，我们可以去改变我们的生活，使它变得更好。这个纪念园为人们提供了多种了解雷迪生平的方式，用不同的形式让人们理解其中的深意。

在整个400 000m²的地块上，纪念园占地5000m²。纪念园的修建是为了让人们永远铭记雷迪博士从住所到实验室的人生之路。纪念园的设计主要借助于树木，有银色橡树大道，高莫哈树小广场，阿育王树大道、棕榈树大道以及木麻黄树柱廊，它们分别象征着"企业家之路""三昧之路""Pradakshna之路""发现之路"以及"慈善之路"。雷迪博士纪念园以自然为背景，在宁静的气氛中与周围自然环境相互联系和交流。

银色橡树大道与企业家之路描绘了雷迪博士一生的轨迹，从一个卑微的人成长到一个成功的企业家。他的自行车与汽车置于企业家之路两端的缓坡上。这条道路可以激励鼓舞人们设定更高的目标，实现更高的目标。

高莫哈树小广场有一处线性水体，其尽头就是三昧之地（译者注：Samadhi，来源于梵语samadhi的音译，意思是止息杂念，使心神平静，是佛教的重要修行方法）。当人们向三昧之地走去，透过墙上的孔洞，可以看到水体映射着天空，人们会感觉到雷迪博士已经离世。雷迪博士曾激励着人们的生活，滋养着人们的生活，而三昧之地中央的留空，代表着一个时代的终结，代表着雷迪博士已经从我们的生活中消失。

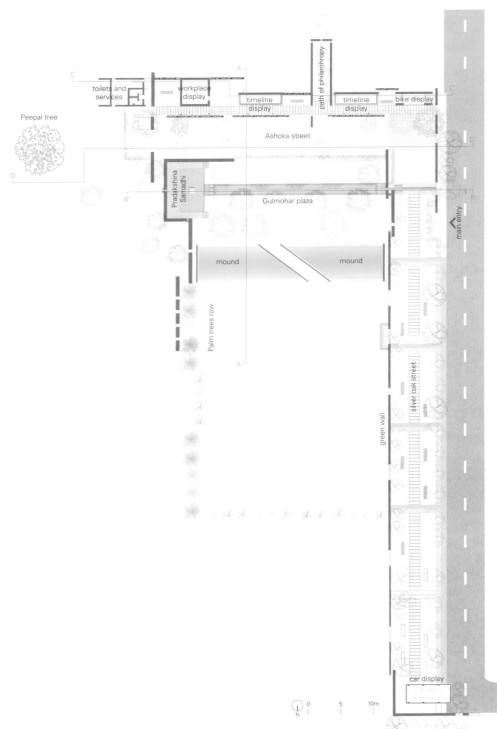

Pradakshna之路围绕三昧之地，与棕榈树大道连为一体。棕榈树之间墙体和空隙象征着雷迪博士的生命从有到无，从自然中孕育，最后又归于自然的过程。空隙数量的逐渐增加，象征着与天空越来越近，最终完全与天空融为一体。

　　发现和启迪之路沿着阿育王树延伸，展示了雷迪博士如何从一个农民的儿子成长为一名企业家，其中充满了艰辛与挑战。铺地材料的材质从最开始的粗糙到半抛光，再到完全抛光，最终与草坪融为一体，以象征着启蒙的菩提树结束。

　　慈善之路则纪念雷迪博士作为慈善家所做出的贡献。

　　从水渠和喷口溢出或涌出的水就像一个祈祷者的自我奉献。雷迪博士一生最重要的一个方面就是渴望回馈社会的愿望。

　　南北轴线上有一条狭小的木麻黄树柱廊。木麻黄树木中掩映着一面石墙，墙上镶嵌着一些展板，向人们展示着雷迪博士的慷慨。

Scientist, entrepreneur and philanthropist, Dr. Kallam Anji Reddy's passion for drug discovery and his pioneering contributions to making medicines affordable are legendary. Born in Tadepalli, India, he worked at the state-owned Indian Drug and Pharmaceuticals for six years and later established Dr. Reddy's Laboratories in 1984.

His life was engaged in ways to alleviate suffering by making affordable medicines, with intense attention to learning and applicating. He stands tall as an example of a deeply compassionate human being who actively tried to rethink the idea of charity and how to unlock potential in the disadvantaged.

The Dr. Reddy Memorial contemplates different aspects of his life with its unique and bold message: no matter how humble our beginnings, it's up to us to transform our lives and reach for something larger than ourselves. This memorial presents multiple ways of navigating Dr. Reddy's life, tracing its patterns to reach an understanding of its lessons. The memorial, occupying 5,000m² out of 400,000m² site,

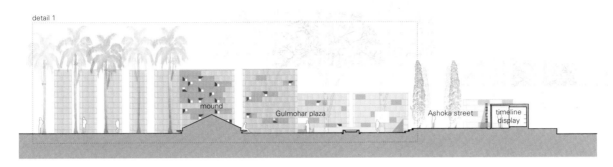

A-A'剖面图 section A-A'

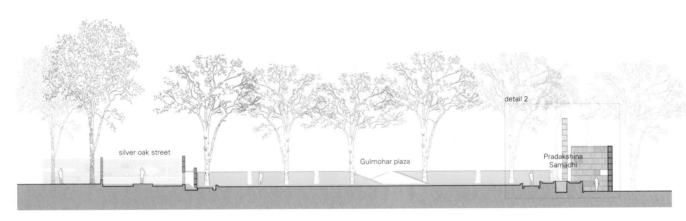

B-B'剖面图 section B-B'

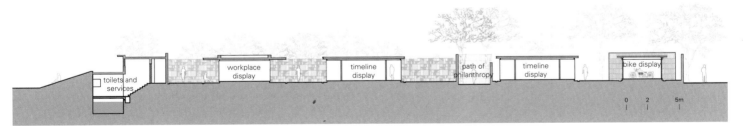

C-C'剖面图 section C-C'

项目名称：Dr. Anji Reddy Memorial / 地点：Hyderabad, India / 建筑师：Mindspace-Sanjay Mohe, Avinash Ankalge
项目管理团队：P. Chandrashekhar Reddy, Ananda Murali Mohan.K. / 平面设计师：Elephant Design-Ashwini Deshpande, Shrish Tilekar
结构顾问：Rays Consulting Engineers / 泳池管道顾问：Astral Pools / 电气与照明顾问：Linus Lopez / 景观建筑师：Design Milieu / 石材覆层与哥哩砖墙：Shri Sai Stones, Sunil Diwakar, Discoy / 客户：Dr. Reddy's Laboratories Ltd., Hyderabad
用地面积：7,450m² / 建筑面积：1,050m² / 总楼面面积：850m² / 设计时间：2014 / 施工时间：2014—2015 / 竣工时间：2016
摄影师：courtesy of the architect-p64, p.66; ©Sebastian Zachariah (courtesy of the architect)-p.58~59, p.60~61, p.62~63, p.68, p.69

详图1 detail 1

was built to immortalize the path taken by Dr. Reddy from his residence to the lab. This place, identified by its trees, became the reference to the design. The avenue of silver oaks, the grid of Gulmohar trees, the avenue of Ashoka trees, the avenue of Palm trees and the colonnade of Casuarina were transformed into the entrepreneurial path, the path to Samadhi, the Pradakshna Path, the path of discovery, and the path of philanthropy, respectively. Dr. Reddy's memorial, set amidst nature, connects and communicates with its natural surroundings in a serene atmosphere.

The avenue of silver oaks and the entrepreneur path, portray Dr. Reddy's life journey from a humble start to a successful enterpriser, with his bike and car displayed on either side of the path with a gradual slope. A walk along this path would inspire one to set and reach higher goals.

A grid of Gulmohar trees with a linear water body comes to an end with Samadhi. As one walks towards the Samadhi, the body of water reflecting the sky is seen through the cut in the wall behind, evoking a sense of his absence. A void in the center of Samadhi represents the end of an era and his absence in the lives he inspired and nurtured.

The Pradakshana path is around the Samadhi along the avenue of Palm trees. The cycle of his life "from" nature and "to" nature is represented through walls and voids between the Palm trees. The number of voids increases, reaching towards and finally merging into the sky.

The path of discovery and enlightenment along the Ashoka trees shows his challenges and growth from a farmer's son to an entrepreneur. The texture of flooring from rough, semi-polished, polished to merging into lawn culminates in the Bodhi tree, a symbol of enlightenment.

The path of philanthropy commemorates Dr. Reddy's contributions as a philanthropist.

The overflowing water channel and spout reach out as an offering of oneself, a prayer. A defining aspect of Dr. Reddy was his desire to give back to his society.

The narrow colonnade of Casuarina lies along the north-south axis. It displays panels engraved on a stone wall along the Casuarina trees to symbolize the generosity of Dr. Reddy.

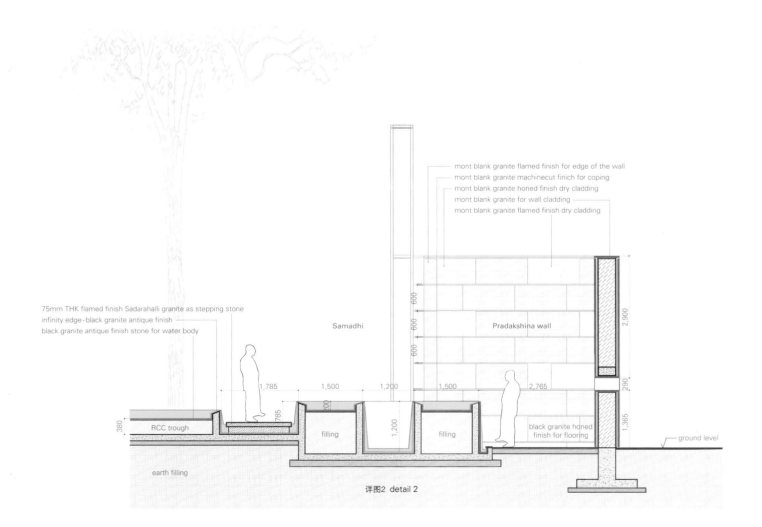

详图2 detail 2

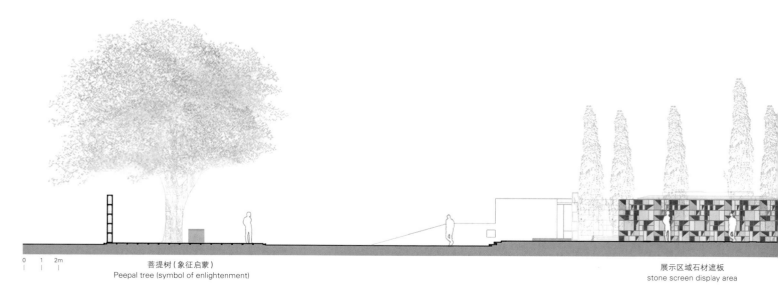

菩提树（象征启蒙）
Peepal tree (symbol of enlightenment)

展示区域石材遮板
stone screen display area

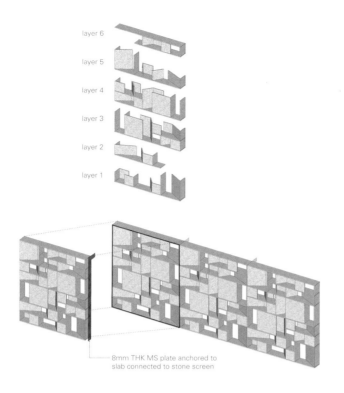

8mm THK MS plate anchored to slab connected to stone screen

展示区域标准石材遮板模块
typical stone screen module for display area

8mm THK MS plate anchored to slab connected to stone screen
MS column clad with ACP sheet
20mm THK mont black machine cut granite stone used for stone screen

display area corridor

20mm THK mont black machine cut granite stone single pieces stuck together with insertion of coin joint between 2 stones
20mm THK honed finish black granite for flooring

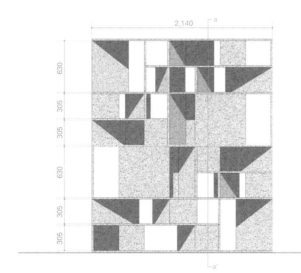

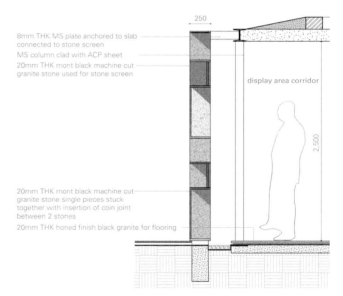

a-a'详图 detail a-a'

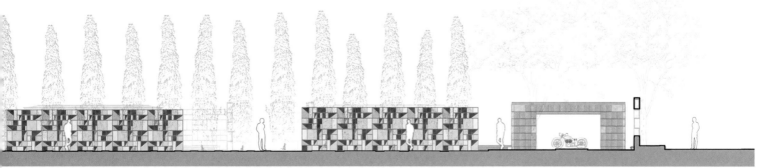

D-D'剖面图 section D-D'

Rajdoot自行车展示区 Rajdoot bike display

Imnang 文化公园
Imnang Culture Park
BCHO Architects

　　该项目是为纪念世界著名钢铁联合企业的奠基人朴泰俊先生而修建的一座纪念馆。除了展示他在事业和慈善方面所取得的成就之外，该纪念中心修建的另一个目的是让人铭记他的个人生活、他的家人以及那些和他共同生活在这个村子里的人们。

　　因此，该建筑就位于他生前曾居住的居民区里。纪念馆所在地被遗赠给伊姆南市，周围三个地块也被纳入规划，占地面积有所扩大。纪念馆由赵炳秀操刀设计，2013年赢得设计竞赛，然后历时两年建造，于最近完工。该市对此最初的设计方案是建造一个引人注目的博物馆综合体来纪念朴泰俊先生的生平和他所取得的成就。这样做需要拆除所有老建筑，只保留朴泰俊先生出生的地方。然而，建筑师的想法与此完全不同。他设计了一座朴素且面积不大的纪念馆，与相邻的三座房屋和两棵大松树一道，安静地缅怀逝者。该纪念馆不仅仅是为纪念一个伟人而修建，项目传达的是朴泰俊在与其他人交往中体现出的温暖和怜悯，其中包括他对居住多年的社区的深切关注。

　　纪念馆将山景引入庭院内，环抱庭院内的两棵大树。这两棵大树，一棵年代悠久，神圣庄严，另一棵非常有名，朴泰俊先生生前常常坐在这棵大树周围，看孩子们嬉戏玩耍。该设计没有拆除位于绵延山坡上的原有建筑，而是将各种原有建筑与新建筑交相辉映，使其焕发新的生机。朴泰俊纪念馆依其所在的起伏不平的地形而建，与原有的周围环境和谐共存，在新老建筑之间营造出独特的对话效果。游客走进纪念馆，一条环路把所有看似分散的元素——树木、原有的房屋等等联系起来，使游客完成一次不间断的体验。

　　游客沿路会看见许多大树，包括"爷爷树"，或可以欣赏到经过精心设计的天空景观。纪念馆中，在原有建筑物和环绕它们的路径之间会出现一些不期而遇的间隙空间，这些空间很难严格地以新旧划分，为我们提供了一个新旧时空可以彼此对话的活力空间。该建筑的空间一方面要体现周围环境的宁静，另一方面又要体现朴泰俊生前个人的生活情形与状态，空间关系交错纷杂。纪念馆内部通过明亮和幽暗空间的对比，使参观者了解朝鲜战争发生之前朴泰俊成长经历的艰辛以及他早年间创业的艰辛，引发参观者对此的思考，甚至同情。

　　走在不断延伸的小径上，参观者可以去感受大地、天空以及周围的环境。纪念馆既安静又幽静，为参观者营造了一个可以沉思的空间，远离外面世界的喧嚣。在这样一个僻静的环境里，参观者或行，或坐，或沉思，所有的烦恼、痛苦和悲伤都会找到属于它们的安放之所。建筑师并没有采用通常的做法，把自然景观平铺直叙地展现在参观者面前，而是提供了让人间接感受自然的方式。当参观者走进庭院，坐在树下，感受到的不是与世隔绝，而是沉浸在周围的社区和丘陵的丰富质感之中。也许，参观者会怀着一颗热情的心和开放的心态来拥抱现实，与朴泰俊先生自己多年来所做的无异。

This project is to create a memorial in honor of Park Tae-joon who founded and ran one of the world's most prominent steel companies. Beyond merely exhibiting his business and philanthropic pursuits, the center's purpose is to remember and to celebrate his personal life, his family and the people with whom he lived together at the village. In lieu of this, the building is located in the typical residential neighbourhood where he resided before his death, at which point it was bequeathed to the city of Yimnang and expanded by incorporating three surrounding lots.

The Memorial Hall, designed by Cho Byeong-soo after winning the competition in 2013, was completed recently after 2 years of construction. The initial plan of the city was to create a prominent museum complex to celebrate the life and achievements of Park Tae-joon in a place where all the old buildings are demolished, leaving his birthplace as it is. However, the architect took a different position. He conceived a small and modest memorial hall, quietly to honor the deceased, by embracing the three neighbouring houses and two large pine trees.

Rather than creating a memorial just about one great person, the project generates the warmth and compassion that connected so many others to him, including a deep concern with the community where he lived for many years.

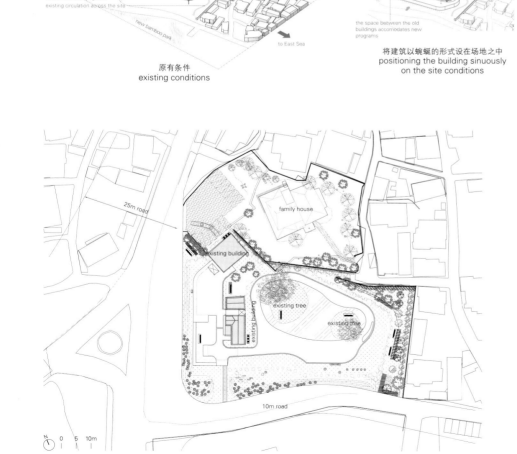

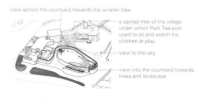

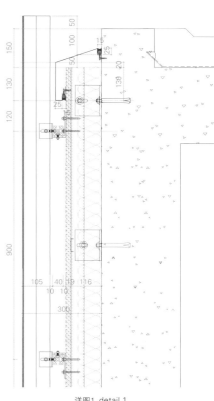

详图1 detail 1

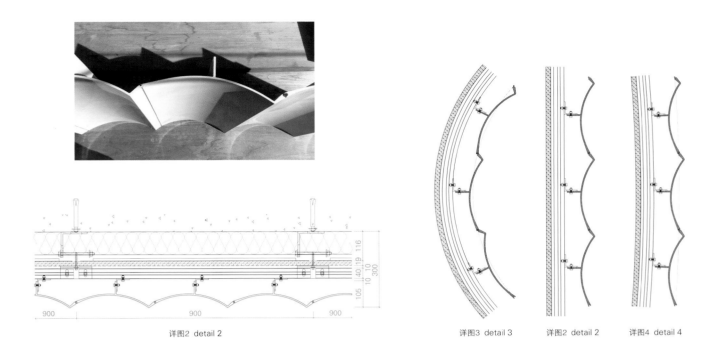

详图2 detail 2

详图3 detail 3　　详图2 detail 2　　详图4 detail 4

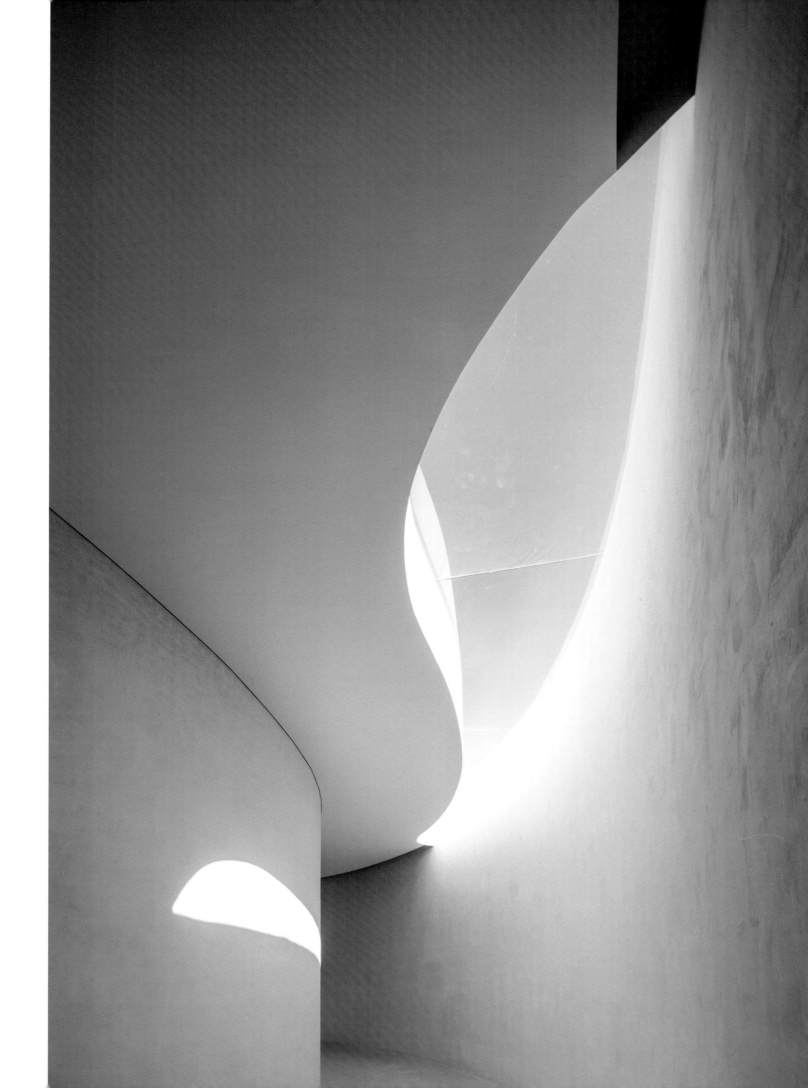

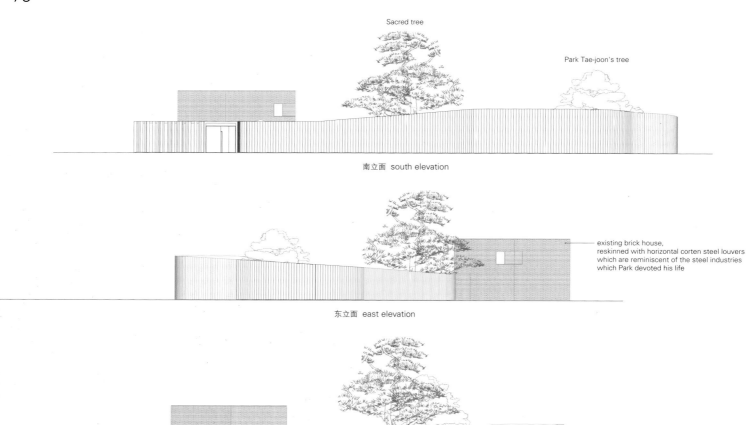

南立面 south elevation

existing brick house, reskinned with horizontal corten steel louvers which are reminiscent of the steel industries which Park devoted his life

东立面 east elevation

西立面 west elevation

1. 大厅 2. 展览空间 3. 室外走廊 4. 图书馆与咖啡厅 5. 办公室 6. 教室 7. 档案室 8. 水池
1. lobby 2. exhibition space 3. exterior corridors 4. library and cafe 5. office 6. education room 7. archive 8. water pond

二层 first floor

屋顶 roof

The Memorial Hall entices the hill by bringing it into view within the courtyard and clasping the site's trees by highlighting the oldest sacred tree as well as another prominent tree around which Park Tae-joon would sit and watch his children play.

Instead of demolishing the tapestry of extant structures along the rolling hillside, the various existing structures are installed on the site with the new structure, bringing them a renewed sense of life and a revitalized understanding through intentional interplay with the new structure. The Park Tae-joon Memorial undulates around its conditions, coexisting with the original surroundings and creating a very distinct conversation through its architecture between old and new.

Once inside, a visitor walks a loop which ties all the seemingly scattered elements – the trees, the existing houses, et cetera – into one uninterrupted experience. Along the continuous walk, one comes across such features as the enormous trees, including the Grandpa Tree, or being presented with carefully choreographed views of the sky.

At the memorial, unforeseen interstitial pockets emerge from between the existing structures and the pathways that encircle them. These are neither strictly new nor old but rather present us with re-enlivened spaces where the two temporal states can converse.

The building's spatial relationship constantly crisscrosses

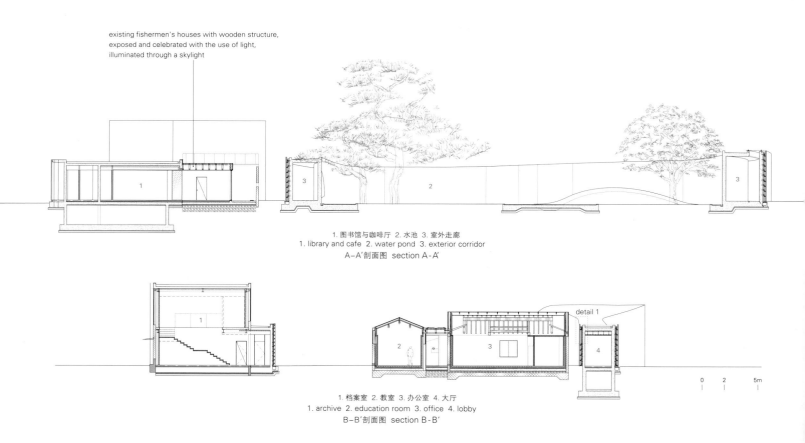

项目名称：Imnang Culture Park / 地点：Imnang-ri, Jangan-eup, Gijang-gun, Busan / 建筑师：BCHO Architects / 项目团队：Kim Sook-jung, Lee Joo-hyoung, Yu Issac, Kim Jae-gi, Hong Kyung-jin _ construction supervision, Choi Dong-uk _ construction supervision / 施工公司：Kyobo Construction / 用途：museum and library
用地面积：4,067m² / 建筑面积：797.35m² / 总楼面面积：952.31m² / 建筑规模：one story below ground, two stories above ground / 结构：reinforced concrete
设计时间：2013.4—2014.12 / 施工时间：2015.6—2017.5 / 竣工时间：2018 / 摄影师：©Sergio Pirrone (courtesy of the architect)

between contemplating on one hand the serene qualities of its context and on the other acknowledging the circumstances of Park Tae-joon's life itself. Contrasting spaces of light and dark invite the visitor to learn and to ponder, if not even to empathize with, the hardships of Park's upbringing before the Korean War and his early business career.

The continuous pathway gently invites the visitor to examine the earth, the sky and surroundings. Both peaceful and secluded, it creates a meditative space for those at the Memorial Hall, isolated from their chaotic lives outside, in a secluded environment where all troubles, anguish and sadness might be eased by carefully understanding each of them on their own terms while walking, sitting and meditating. Curated views of nature focus less on showcasing it a point blank range as is often the case but, alternatively, on providing ways to feel nature indirectly.

When the visitor enters the courtyard and sits beneath the trees, he might feel not so much as if in a detached state of being but rather immersed in the rich texture of the surrounding neighbourhood and hills. Perhaps the visitor might come to embrace the realities with a warm heart and an open mind, not unlike how Park Tae-joon himself did for so many years.

Len Lye中心是新西兰唯一一座为一位艺术家专门修建的博物馆，其设计深受一位艺术家的生活、思想、著作和工作的影响。这位艺术家就是Len Lye（1901年7月5日—1980年5月15日）。Len Lye曾在1964年说过："伟大的建筑也是伟大的艺术。"这句话被Patterson联合建筑师事务所奉为圭臬。收藏Len Lye作品的Antipodean寺庙的设计方法和建筑形式都受此影响。

Lye对寺庙情有独钟。在构思该建筑的整体设计时，建筑师从古典建筑中的"megarons"（正厅），也被称作大堂（译者注：古希腊和小亚细亚建筑的中央部分，即中央大厅），以及从波利尼西亚建筑风格的形式和设计理念中汲取灵感，无论从历史的角度还是美学的角度来说，都是自然而然的。这些也曾影响Lye，毕竟Lye自己就是客户。

为了用新的方式做到这点，Patterson联合建筑师事务所的建筑师们使用"系统方法论"，即一种整体或自适应的方法进行研发设计。这意味着在设计中不是采用比例或是美学的方法，而是在项目的生态环境中使用模式来驱动设计元素。例如，采用当地生产的不锈钢来制造出波纹般、光亮闪闪的柱廊外立面，将Lye在动力学和光线方面的创新与该地区的工业创新联系在一起。通过这样创意的结合，建筑师们将Lye的作品作为对塔拉纳基（译者注：新西兰的一个区）的献礼。

柱廊波纹般的外立面就像是一块电影幕布，有三个不对称的倾斜侧面，形成一种门廊类型的空间，在古希腊中被称为pronaos（门廊）。柱廊围合而成的空间是一个画廊，用来珍藏和展示Lye的大型作品。从空中向下看，柱廊的顶部边缘形成一朵银蕨花（译者注：新西兰的国花）图案，展示了该博物馆设计深受对Len Lye有很大影响的波利尼西亚风的影响。柱廊排列形成一个大的门廊，使主画廊如同"megaron"，其作用如同wharenui（译者注：毛利会堂）。这些神灵和祖先都是Lye那给人灵感和启迪的作品中经常借鉴参考和表现的。围绕在周围的每一根重复的14m高巨大石柱都是由一个混凝土预制构件建成的。

一般来说，寺庙中最神圣不可侵犯的地方"adyton"（阿底顿，阿波罗神庙中的密室）位于离入口最远处。在这栋建筑中，这样的位置存放着关于Len Lye的所有档案文件。而博物馆的"宝库"，也就是所谓的"opisthodomos"（译者注：后室，庙宇内用作珍藏室的内殿小室），则位于博物馆入口处的上方，俯瞰着下方进出来往的游客。

该项目在充分尊重原有较小的Govett Brewster美术馆的基础上，与之连为一体。Govett Brewster美术馆是由该市的一所废弃不用的电影院翻新而成的，属于受保护性文物建筑。博物馆和美术馆彼此相连。在这样一座共用的灵活建筑中，游客通过一条环形通道来欣赏博物馆和美术馆所展出的丰富多样的展品。

在环形通道上，光线通过柱廊的光阑被引入室内，在通道上形成移动的光线图案。所以说，该建筑也许可以说是一种被动的动态建筑。建筑师们希望该设计能够挑战纯粹现代主义在当代思想中的主导地位。古典主义已经过时了几十年，Len Lye博物馆试图使现代主义语言更富意义。与包豪斯的传统相比，该博物馆的空间更清晰透明，更宏伟成功，更欢快喜庆，同时也比轴式建筑更有说服力，更流畅。

The Len Lye Center is New Zealand's only single artist museum and its design is deeply influenced by the life, ideas, writings and work of Len Lye (5 July, 1901 – 15 May, 1980). It was Lye himself who said in 1964 that "great architecture goes fifty-fifty with great art", a maxim that has informed the approach and the form of the Patterson Associates-designed Antipodean Temple that houses his work.
Lye was fascinated with temples and in conceiving the overall design it seemed aesthetically and historically appropriate to draw inspiration from the "megarons", or great halls, of the classical world, as well as Polynesian forms and ideas. These also influenced Lye and he was, after all, the client. To do this in a new way, Patterson Associates' design was developed in a holistic or adaptive way, using "systems methodology". This means that rather than using proportion or aesthetics, they use patterns in the ecology of the

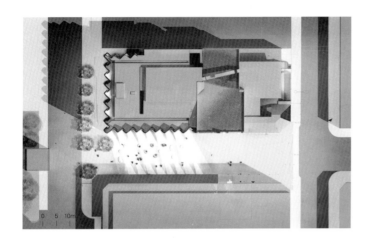

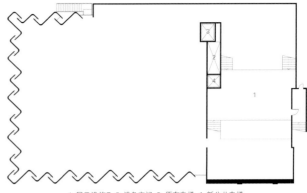

1. 展示设施C 2. 设备空间 3. 原有电梯 4. 新公共电梯
1. display service C 2. services 3. existing lift 4. new public lift
四层 third floor

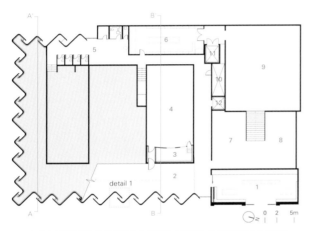

1. 入口大厅/接待处 2. 露台走廊 3. 放映室 4. 电影院 5. 卫生间门厅 6. 档案馆
7. 画廊1 8. 画廊2 9. 后台 10. 设备空间 11. 原有楼梯 12. 新公共电梯
1. entry lobby/reception 2. terraces gallery/foyer 3. projection room
4. cinema 5. W/C lobby 6. archives 7. gallery 1 8. gallery 2
9. back of house 10. services 11. existing lift 12. new public lift
一层 ground floor

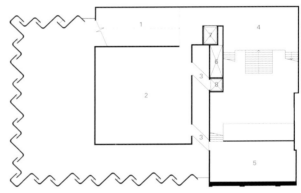

1. 上升坡道 2. 主画廊/可细分空间 3. 连接桥 4. 展示空间A 5. 展览空间B
6. 设备空间 7. 原有电梯 8. 新公共电梯
1. ramp up gallery 2. main gallery/sub divisible 3. bridge 4. display space A
5. display space B 6. services 7. existing lift 8. new public lift
三层 second floor

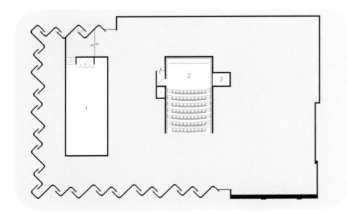

1. 电动机室 2. 电影院 3. 钢琴存放室
1. motor room 2. cinema 3. piano store
地下一层 first floor below ground

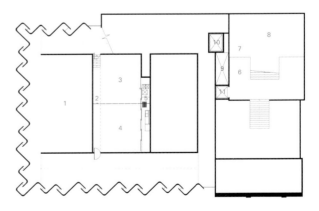

1. 大型艺术品画廊 2. 上空空间/采光井 3. 教育工作室A 4. 教育工作室B 5. 工作室湿区
6. NZ电影档案室 7. 纸质作品储藏室 8. 后台 9. 设备空间 10. 原有电梯 11. 新公共电梯
1. large works gallery 2. void/light shaft 3. education studio A 4. education studio B
5. wet area for studio 6. NZ film archive 7. storage of paper works
8. back of house 9. services 10. existing lift 11. new public lift
二层 first floor

90

项目名称：Len Lye Center / 地点：New Plymouth, New Zealand
建筑师：Pattersons / 设计主管：Andrew Patterson
项目主管：Andrew Mitchell / 项目团队：Andrew Patterson, Andrew Mitchell, Daniel Zhu, Caleb Green, Joanna Aikens / 施工：Cleland Construction
结构工程师：Holmes Consulting Group
机电工程师：E-Cubed Building Workshop
工料测量师：Rider Levett Bucknall / 立面工程师：Mott MacDonald
声学工程师：Marshall Day / 防火工程师：Holmes Fire
电影院地毯：Irvine Flooring / 细木工：IMO
客户：New Plymouth District Council
用地面积：3,600m² / 总楼面面积：2,000m²
室外饰面：No. 8 polished stainless steel, glass & precast concrete panels
室内饰面：exposed concrete & white plasterboard to gallery walls and ceilings
屋面材料：silver veedek diamond roofing & sikaroof MTC membrane
地板材料：light oxide concrete
窗户材料：structural glazing (performance specified) by Mott MacDonald
项目时间：2010—2015 / 竣工时间：2015
摄影师：©Patrick Reynolds (courtesy of the architect) - p.84~85, p.86, p.87, p.88, p.90, p.91
©Sam Harnett (courtesy of the architect) - p.93 left, right-middle, right-bottom
©Bryan James (courtesy of the architect) - p.93 right-top

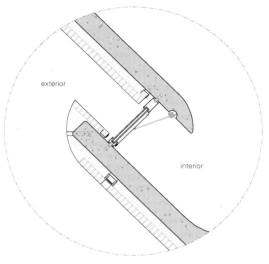

详图1 detail 1

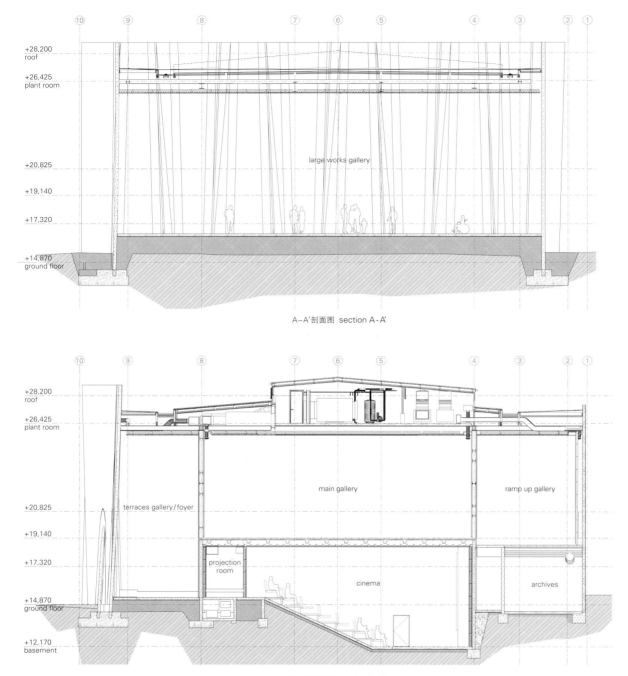

A-A'剖面图 section A-A'

B-B'剖面图 section B-B'

project's environments to drive the design elements. For example, the shimmering, iridescent colonnade facade, manufactured locally using stainless steel links both Lye's innovations in kinetics and light as well as the region's industrial innovation. By doing this the architects celebrate the fortunate gift of his works to Taranaki.

The colonnade creates a theatre curtain, but with three asymmetric ramped sides, leading to a type of vestibule, known as "pronaos" in Ancient Greece. This is formed by the gallery holding the large Lye works. Viewed from above, the colonnade's top edges create a koru form, displaying the Museum's Polynesian influences as the meeting house, or wharenui, for Len Lye. The procession of the colonnade morphs into a portico, announcing the main gallery as a type of "megaron" but also functioning as a wharenui; the deities and ancestors were referenced and represented by Lye's inspirational work. Each of the 14-meter-high monolithic repeating columns surrounding it is constructed of a single precast piece of concrete.

Traditionally, the most sacred and private part of a temple,

the "adyton", is located at the point furthest from the entrance. Here is housed the Len Lye archive, while the "treasury", known as the "opisthodomos", looks back to the people entering below.

The project respectfully links into the smaller existing Govett Brewster Art Gallery, which itself has been retrofitted from the city's decommissioned heritage cinema. The combined facility is undivided, with a circular loop allowing visitors to appreciate the changing museum and gallery displays within one flexible and shared structure.

On the circular loop, light is drawn inside through the apertures in the colonnade, and these create moving light patterns on the walkway, perhaps a form of passive kinetic architecture. The architects hope the design challenges the dominance of pure modernism in contemporary thought. Classicism has been unfashionable for many decades and the Len Lye Museum seeks to extend modernist language with meaning. The Museum space is more lucid, triumphant and celebratory than Bauhaus traditions, but also more cogent and flowing than axis-generated architecture.

Memory not Perfect

记忆并非完美

记忆之家_House of Memory / baukuh
阿尔托大学主楼Dipoli大楼_Dipoli, Aalto University Main Building / ALA Architects
海法大学Younes & Soraya Nazarian图书馆_University of Haifa, Younes & Soraya Nazarian Library / A. Lerman Architects
少女塔的修复_Restoration of the Maiden Tower / De Smet Vermeulen Architecten + Studio Roma
安吉·雷迪博士纪念园_Dr. Anji Reddy Memorial / Mindspace
Imnang文化公园_Imnang Culture Park / BCHO Architects
Len Lye中心_Len Lye Center / Pattersons
于特岛赫恩胡塞特纪念馆和学习中心_Hegnhuset, Memorial and Learning Centre in Utøya / Blakstad Haffner Architects
科泰艺术展馆_KOHTEI / SANDWICH
马科瓦的Ulma家庭博物馆_The Ulma Family Museum in Markowa / Nizio Design International
卡廷博物馆_Katyń Museum / BBGK Architekci
大地的见证——侵华日军第七三一部队罪证陈列馆
The Witness of Land – Memorial Hall of Crime Evidences by Unit 731
_ Architectural Design & Research Institute of SCUT South China University of Technology
赫茨尔山阵亡士兵纪念馆_Mount Herzl Memorial Hall / Kimmel Eshkolot Architects
渥太华国家大屠杀纪念碑_National Holocaust Monument Ottawa / Studio Libeskind
第二次世界大战博物馆_The Second World War Museum / Studio Architektoniczne Kwadrat

记忆并非完美_Memory's not Perfect / Diego Terna

以记忆为主题的设计项目面对的是由个人记忆组成的世界,游客的个人记忆可以通过他们所在的同一空间得到激活。在建筑中,这是一个基本的主题,它在现在和过去的对话中,通过纯粹的建筑手法或者更隐喻和更具象征性的设计,已然确立了良好的存在感。

不论形式如何,激活个人对过去的理解对于通过建筑空间来定义合理未来这件事情十分重要。这可以是一场探索之旅,像电影《记忆碎片》里面的主角所经历的;也可以是传记式的概述,如阿尔多·罗西的设计那样;它还受到原有环境的约束,如里昂·巴蒂斯塔·阿尔伯蒂的设计作品。虽然还不十分确认,但过去已建立起自己的记忆图集,并将营造出空间叙事故事。

记忆和过去为以下我们将要分析的设计项目奠定了基础,但它们是灵活的元素,只是从过去汲取力量而没有拘泥于过去的形式,从而使自身融入创新的未来。

Projects that work on the theme of memory face a universe made of personal remembrances, activated in the visitors through the same space in which they are located. It is a constitutive subject of architecture, which has established a good part of its existence on this dialogue between present and past, through a purely architectural operation or through a more metaphorical and symbolic work.

Whatever the modality, activating a personal interpretation of the past is fundamental for defining a plausible future through the built space. It is a journey of discovery, like the protagonist of the movie *Memento*; it can be a biographical survey, as in Aldo Rossi's projects; it can also be almost obliged by an existing situation, as in the works of Leon Battista Alberti. Without certainties, but constructing an own atlas of memories, past will structure the narration of space.

Memory and past build a foundational basis for projects that will be analyzed below, but they become living elements only thanks to a vision that is not anchored to this past, but that draws strength from it, for submerging, then, in an innovative future.

记忆并非完美

Diego Terna

我们谈到建筑时,记忆主题是基础性话题:建筑从一开始在某种程度上就与记忆的概念直接相关。就是说,如果建造一个空间的第一步是要尝试改变人们生活的环境条件,那么建造建筑也是一样,因为建筑也是被"建造"起来的,这是一个始终存在的因素,对于后人来说,建筑不仅会界定一个地方,还能创造一片环境和一种情境。

尽管建筑只是暂时的,但从其属性和概念来看,相比于空间,它注定更具物理优势。只要它能被保留下来,便会始终存在。在这个意义上,我们有必要运用记忆、过去或与之接近的手段,去继续寻找其存在的历史意义,它将带领我们回到过去,帮助我们建造一个新的未来。

记忆也是对个体的一种挑战,比如构筑一个仅仅保留在个人的回忆之中,但现实却并不存在的画面。就像用力一点一点地刷洗一堵墙,然后深挖进里面去寻找回忆,找到的只是那一个可能并不是真的而只是虚拟的画面。这样是会出问题的。不论是拆解记忆还是重构记忆,都是一件很难的事情,正是这个难度导致电影《记忆碎片》的主角莱纳德·谢尔比宣布其经过思考得出如下结论:"记忆并非完美,记忆甚至是不美好的。问一问警察就知道,目击者的证词是否绝对可信。警察光靠坐着回忆事情是抓不到杀人犯的。他们得收集事实证据,做笔记,推导结论。破案要靠事实,而不是记忆。……看,记忆中一个房间的形状会变,一部汽车的颜色会变,记忆甚至可能会被扭曲。它们只能作为一种解释而存在,而不能当作事实记录。一旦你手上有了事实记录,记忆便与你不相干了。"

该影片导演克里斯托弗·诺兰在整部电影中努力去解释这场以部分记忆为基础,实为事实正名的研究:他打造了一些连续的幻象

The theme of memory is a foundational topic when it comes to architecture: the very origin of architecture is in some way linked to the concept of memory. If the first act that constructs a space is the attempt to change the environmental conditions in which people live, the same construction of architecture finds the main point of its narration in the fact that it is "constructed", that is an element that will remain in the history, which will define a place, an environment, a situation for the future generations.

Although temporary, architecture is destined to take the physical advantage over the space, by its nature and conception: it remains, it stays. In this sense it has a continuous need to seek the historical meaning of its existence with a work on memory, on a past, more or less close, that returns and determines a new type of future.

Memory is a challenge to oneself, to the construction of the image of a past that doesn't exist and it has remained fixed only in personal remembrances. This can be problematic, a wall to be scrubbed bit by bit, digging into memories, in an imaginary that, perhaps, is not real, but only invented. It is this difficulty of de-construction and re-construction of memories that induces the protagonist of the movie *Memento*, Leonard Shelby, to pronounce this reflection: *Memory's not perfect. It's not even that good. Ask the police. Eyewitness testimony is unreliable. The cops don't catch a killer by sitting around remembering stuff. They collect facts, they make notes and they draw conclusions. Facts, not memory. [...] Look, memory can change the shape of a room. It can change the colour of a car and memories can be distorted. They're just an interpretation. They're not a record. They're irrelevant if you have the facts.*

The director Christopher Nolan tries to explain this research for the truth, based on partial memory, throughout the movie: he builds continuous simulacra and evidences in support of facts that, perhaps, do not exist; often,

Malatestiano寺庙，意大利里米尼
Tempio Malatestiano, Rimini, Italy

和证据以支撑事实，但这些东西也许都是不存在的，甚至在大多数情况下，它们不过是一个健忘的脑袋里的虚幻重构。然而正是这些难以解释的疑点成就了一个极富想象力的复杂影片，就像影片中那些复杂的展开过程一样，能够在记忆最黑暗的褶皱中发现自己，并明确主人公头脑的建设性作用。

那些记忆都是很私人的，或许不那么真实，但可作为重构接近真相的必然结果，亦可作为空间创新的元素被转置到建筑之中。这些正是阿尔多·罗西传记中记述的，阿尔多在厄尔巴岛的小屋里发现了一种典型的住宅原型，该原型利用祖先留下的记忆（四面墙、一个三角形屋顶，就像小孩子设计的一样）提出空间进化的概念："在我的设计中，沙滩小屋是一再出现的主题，因为我总能在其间发现一种建筑的综合运用或简化。我喜欢把它们变成室内景观。"

厄尔巴岛的小屋正是一种建筑观点的表达，是对个人回忆的一种掌控，这点在这位意大利建筑师打造的连续空间中有所体现：此处所说的记忆，是一种平息建筑形状讨论的投机取巧，或者说是在设计中使形状合法化的权宜之计。多亏了它，罗西通过人们记忆中可识别的和熟悉的几何图形，找到了一个可以帮助人与建筑直接进行沟通的方式。

但是人与建筑之间的对话也可能发生暂时性分层，这就承认了过去的记忆是仅作为项目元素之一而存在的，它是建造空间的起点。

这是一种设计态度，长久以来，它一直标志着当代建筑的发展之路，在阿尔多·罗西的设计中，这也是一种基础的设计灵感。从很早以前，这种态度就已渗透到建筑史之中，在文艺复兴时期最大限度地展现了出来（此时经典的希腊和罗马风格被再次发掘出来）。

they prove to be only fallacious reconstructions of a forgetful mind. Yet these difficulties lead to a fertile complexity, as observed in the intricate unfolding of the events of the film, capable of unravel itself within the darkest folds of remembrances and explicit the constructive role of the mind of the protagonist.

The memories, very personal, perhaps untrue, certainly result of reconstructive approximations, can become, once again transposed into architecture, elements of spatial innovation: this is what happens in the biographical evocations of Aldo Rossi, who finds in the huts of the Elba island a sort of iconic residential prototype, which exploits an ancestral remembrance (4 walls, a triangular roof, as designed by a child) to propose an evolution of space: *Beach huts are a recurring theme in my designs because I have always found a sort of synthesis or reduction of architecture in them. I like the idea of them becoming an interior landscape.*

The Elba huts are an idea of architecture, the manipulation of a very personal past, which is then found in the successive spaces of the Italian architect: the memory, here, is the picklock used to disrupt the discourse about the forms, or, better, it is the expedient that gives forms a legitimation within the project. Thanks to this expedient, Rossi finds a very direct way of communicating architecture to people, through recognizable and familiar geometries, which can be found in the memory of people.

But the dialogue between people and architecture can also take place on temporal stratifications, which admit the memory of the past as an element of the project, a starting point for developing the space.

It is an attitude that for a long time has marked the path of contemporary architecture, finding in Aldo Rossi a fundamental inspiration. It has permeated the history of architecture from the ancient times, finding a point of maximum

侵华日军第七三一部队罪证陈列馆，中国
Memorial Hall of Crime Evidences by Unit 731, China

赫恩胡塞特纪念馆和学习中心，挪威于特岛
Hegnhuset, Memorial and Learning Centre in Utøya, Norway

  从这个意义上讲，里昂·巴蒂斯塔·阿尔伯蒂设计的里米尼马拉泰斯塔诺教堂（1450年建）就是一个与历史预先存在有关的例子，它也与记忆的实质性概念息息相关。这座新建筑围绕一座1200年的教堂而建，它为这座古老的建筑竖起了保护栏，但却忽略了老教堂原始的结构。新建筑采用了不同的建造结构，包括一些珍贵的材料，它所采用的建筑语言更多的是与罗马建筑（圆拱的连续重复）产生对话，而不是与老教堂的罗马风格产生对话，不过它设法利用新墙的深度与老建筑交融在了一起。这个深度很好地隐藏了砖块隔墙，但人们还是可以瞥见老建筑的大部分结构：由于建筑结构的不同，这个深度更加突出了设计的差异。简而言之，建筑设计使过去和现在以两种截然不同的语言沟通，但它也同时让建筑或丰满，或复杂，或意外百出。在建筑中，人们的所看到的景象绝不会一成不变，但是这样的设计也会使关于建筑的传说处于危机之中。

  阿尔伯蒂定义了他个人的记忆，对过去的回忆进行了重构，既有最近的记忆（1200年的教堂），又有最远的记忆（罗马古典主义）。记忆的主人公在找寻事实，他们认为记忆是靠不住的，而意大利建筑学家认为不同语言、个人记忆之间的对话与翻译不一定是真实的，但却是似是而非的。

  我们可以在华南理工大学建筑设计研究院设计的哈尔滨侵华日军第七三一部队罪证陈列馆（154页）和布莱克斯塔德·哈夫纳建筑师事务所设计的赫恩胡塞特纪念馆和学习中心（104页）再次发现这种设计态度，新建筑与老建筑融合在一起，，不仅是对老建筑的保护，更是为了彰显记忆。它们被圈在一个空间之中，更加密集，聚拢，且有力量。哈尔滨的罪证陈列馆通过强烈的存在感搭建了一条通往

aspiration during the Renaissance (when the Greek and Roman classical period had been rediscovered). In this sense, the project of Leon Battista Alberti for the Tempio Malatestiano in Rimini (1450) is a fundamental example of relationship with historical pre-existences and therefore with a substantial concept of memory: the new construction is built around a church of 1200, erecting a sort of protective casket for the ancient building, but almost neglecting its original structure. The new church adopts a different structural pace, precious materials, a language that talks more with Roman architecture (the continuous repetition of the round arch) than with the Romanesque of the old church: yet it manages to involve the ancient building through the depth of the new walls. This depth hides the brick partitions, but it allows glimpses of large parts of the old building: it underlines, thanks to the different pace of the structure, the dissimilarities. In short, it lets past and present to speak two apparently different languages, but in doing so, it allows the development of fertile, complex, accidents, in which the views are never banal, but they put into crisis the tale of the architecture.

Alberti defines his personal memory, reconstructing a remembrance of the past, both the closest one (the church of 1200) and the farthest one (the Roman classicism). Where the protagonist of Memento was looking for facts, thinking that memory was fallacious, the Italian architect proposes interpretations, dialogues between different languages, personal memories, not true, but plausible.

We can rediscover this attitude in the Memorial Hall of Crime Evidences by Unit 731 in Harbin by Architectural Design & Research Institute of SCUT(p.154) and in the Hegnhuset Memorial and Learning Center by Blakstad Haffner Architects (p.104), where architecture is incorporated within the buildings: in this case not to protect them, physi-

阿尔托大学主楼Dipoli大楼,芬兰
Dipoli, Aalto University Main Building, Finland

少女塔,比利时
The Maiden Tower, Belgium

记忆的道路。在赫思胡塞特纪念馆,你只能看到一部分老建筑,想要获取完整的记忆是一件不太可能的事情,但是将一部分对外展现,让更多的人了解是十分有必要的。

因此,新建筑与历史建筑的关系,往往能够通过前者对后者的保护体现出来,但这种关系通常将自己定义为现代与过去的衔接点。如ALA建筑师事务所设计的阿尔托大学主楼Dipoli大楼(24页)、A.列尔曼建筑师事务所设计的海法大学图书馆(36页)、De Smet Vermeulen建筑事务所和Roma工作室设计的少女塔(48页),记忆的概念都是通过预先存在来协调的,这种预先存在仍然可以被利用起来,而不是只能远观。

所以,我们需要通过加建和历史分层来定义新与旧的关系、当代与过去的关系,正如阿尔伯蒂设计的教堂,就是一种建筑的升级,它通常需要人们对空间认真且深入地反复思考。

新建筑的构建可以通过最小化和最为集中的干预方式实现,如ALA建筑师事务所的项目;或者通过明显的加建实现,如A.勒尔曼建筑师事务所的项目。这两个项目都通过巧妙加建与修改,与老建筑形成相对直接的对话,将一股清新之气注入古老的空间,并在材料和体量中展现无遗。即使我们观察出更多的相关变化,新的体量也能试图与已经存在的建筑总体设计对接,就像为了完成一幢尚未竣工的遗留老建筑一样。在这种情况下,因为建筑的外形尽量保留了原有的几何形状,那么说到底,建筑材料才是新建筑与老建筑之间差异的决定因素。

cally, but to make the memory more evident. They are enclosed within a space, forced to become denser, more focused, more powerful. In Harbin the architecture constructs a sort of access path to memory, with a totalizing presence; in the Hegnhuset Memorial the project acquires only a part of the old building: it is impossible to contain the entire memory, but it is necessary to leave a part of it outside, in the world, to release it towards a wider knowledge.

The relationship with a historical element, therefore, can occur through its protection, but often defines itself as a counterpoint between modernity and past: in the projects Dipoli, Aalto University Main Building by ALA Architects (p.24), Haifa University Library by A. Lerman Architects (p.36), The Maiden Tower by De Smet Vermeulen Architecten and Studio Roma (p.48), the concept of memory is mediated by a pre-existence that can still be "used" and not only visited from outside.

It is therefore a matter of defining the relationship between new and old, between contemporaneity and past, through an operation of addition, of historical stratification: as in the church of Alberti, an architectural upgrade often requires a necessary and deep rethinking of the space.

It can be done through minimal and focused interventions, as in the ALA's project, or with a more pronounced addition, as in the project by A. Lerman. They are subtle additions and modifications, which build a rather bare dialogue, filling the ancient space with a new air, expressed in the materials and in the volumes. Even when we observe more relevant changes, the new volumes try to adapt to an already existing overall design, as if to complete the non-finish left by the old building. In this case it is the materials, above all, that determine the difference

Len Lye中心，新西兰
Len Lye Centre, New Zealand

　　在少女塔项目中情况也是如此，加建游戏变得更加精妙，它让人们对这种在原有建筑上加建的新元素有了双重解读：在这里，当代建筑以一种模棱两可的姿态吸收了老建筑的元素，建筑仍然采用自己原来的形态，但因为新建筑材料与老建筑材料截然不同，因此建筑变得更为抽象与荒凉，几乎认不出来了。

　　在这个项目中，记忆会引领人们走向一个几乎矛盾的情节：如果新建筑是老建筑的翻译和解释，那么建筑就是一个仁慈的两面派，它不仅向访客们提供一个近乎教诲式的规条，又为老建筑的改变提供了结构支撑。

　　迄今为止，以单纯的新旧建筑对比为基础，所有研究项目都给"记忆"一词赋予了一种深意，正是这种新旧对话造就了空间的复杂性、机遇性，也提供了出人意料的视野和短暂的分层。

　　不过还有一些其他建筑项目，也围绕记忆概念设计，但这里的记忆常常是独立于建筑之外的，通常来说，指的是个人或者历史事件的记忆。因此建筑常常被建造成对现实的隐喻或象征，那些也正是它们的外在延伸，它们算不上是空间支撑，而是为搭建建筑之路提供基础叙述的。

　　因此，这些例子表明，记忆并不是因老建筑的存在而产生的动机，而是回忆中的一页，它必须成为空间，就像阿尔多·罗西·艾尔巴的小木屋，关于记忆的故事往往比现实和真相更有意义，它同时还能够定义建筑物本身发展变化的某一情节。

　　在这些风格混杂的设计项目中，我们可以观察到一些共有的元素：

　　— 墙，一个含蓄的可令人深思的元素；

between the new and the old, since the shapes try to respect the existing geometries.

It happens in the Maiden Tower, where the game of additions becomes more nuanced and allows a double reading of the new signs that add up to the existing building: here the contemporary absorbs the ancient with an ambiguously mimetic attitude; it returns the forms, its own forms, but abstract, gaunt, almost unrecognizable, because the new materials are completely different from the old ones.

In this case, the memory leads to an almost paradoxical work of fiction: the building is a benevolent Two-Face, which is offered to visitors in an almost didactic manner, as if the new building were to be interpretation and explanation of the ancient, but also structural support for its transience.

All the projects observed so far define a sense to the word Memory, on the basis of a purely architectural comparison between the old and the new: it is this dialogue that builds the complexity of the spaces, its accidents, the unexpected views and the temporal stratifications.

Other projects, however, work on a concept of memory that is outside the field of architecture, being, in general, the memorials of people or historical events: the buildings are therefore constituted as metaphors or symbols of realities that are external to them, which do not offer a spatial support but provide basic narratives to develop architectural paths.

In these cases, therefore, memory is not an incentive given by the presence of an ancient building, but a page of remembrances that must become space: as in Aldo Rossi Elba huts, the tale of memory is worth more than reality and facts and it allows to define a plot of development for the architecture itself.

记忆之家，意大利
House of Memory, Italy

Ulma家庭博物馆，波兰马科瓦
The Ulma Family Museum in Markowa, Poland

— 围墙，一个空间管理者；
— 通道，一个冥想主题；
— 光，一个基础布景效果。

这些都属于建筑领域的基本组成元素。它们构成了建筑物的基础，但是在这里，它们也可以有效地被当作某一主题的意向申明：记忆，在这里，连接的是功能与形式。尽管这些元素能够在所有设计项目中找到，但是其中某几个元素往往能够发展成为一种显著特征。

在纪念馆中，墙是语言的栖息地，是解释的平台，它是一个有训导作用，并能够呈现历史和主题复杂性的元素。正如阿尔伯蒂设计的教堂，墙是通过可触摸的庞大的有形支撑来展现记忆的地方。

在帕特森建筑事务所设计的Len Lye中心（84页），有一面如浪花般摆动的镜帘，它能反射出来访者和他们身旁的一切，使人们看到的景象失真，因此人与建筑之间能够产生互动。在baukuh建筑师事务所设计的记忆之家（14页）中，墙造就了一个整洁、含蓄的空间，并能够通过微妙的纹理游戏同周边环境呼应：砖块变成了像素，构成无名的米兰尼斯人物像，远观之时，完整地映入眼帘，效果甚好。Nizio Design International事务所打造的马科瓦Ulma家庭博物馆（132页）的墙以拥抱的形式欢迎访客的到来，同时建筑中容纳了博物馆，不但扩大了室内空间使其延伸到室外，还使建筑空间与连续的景观运动融为一体。外围的空间是一个适于冥想的地方，也是放大回忆以反思的地方，因为它将人们聚集在一个固定的空间中，人们关注的仅是自己本身。用BCHO建筑师事务所的话说：显而易见，韩国式世界观是关于"拥抱"的，那是一种将既有条件、地点和周围环境包裹在一起的"拥抱"。近距离观看之下，我们能够重新发现建筑的意

Some common elements can be observed within this heterogeneous group of projects:
- the wall as a connotative and contemplative element;
- the enclosure as a space organizer;
- the path as a meditative theme;
- the light as a fundamental scenic effect.

These are elements that are generally part of the architectural field. They constitute its foundations, but, in this case, they can be effectively observed as declarations of intent toward a theme: the memory, here, links together function and form. Even if these elements can be found in all the projects, some of these tend to develop a feature more than the others.

In a memorial the wall is the place of the language, of the explanation: it is the didactic element, which can take on the history and complexity of the theme: as in the church of Alberti, it is the place where the memory is revealed, through the physical support of a tangible mass.

In the Len Lye Center by Pattersons (p.84), it is a mirror curtain, moving like a wave, which portrays visitors and their surroundings. It reflects light and images and distorts the views, thus creating an interaction between people and architecture; in the House of Memory by baukuh (p.14), the wall creates a tidy, introverted box, which speaks to the surroundings through a very subtle texture game: the bricks become the pixels that create anonymous Milanese portraits, pushing the views far away, to allow the overall vision of the images; in the Ulma Family Museum in Markowa by Nizio Design International (p.132), the wall welcomes visitors with a hug and, at the same time, the

Imnang文化公园，韩国
Imnang Culture Park, Korea

赫茨尔山阵亡士兵山纪念馆，以色列
Mount Herzl Memorial Hall, Israel

义，而在拥抱之下，在保护之下，人们能够更好地看向外界和外界现实。

在Imnang文化公园里，由BCHO建筑事务所打造的泰俊公园纪念馆（70页）就是由一个抽象空间围合起来的项目，这个抽象空间充满自然气息，为的是尽可能地减少外界的干扰因素，但是它定义的外墙纹理创造了一种连续的重复元素，，如同一种咒语。它的建造契合地形，不但内部空间视野良好，同时还能欣赏到外面的景色。它与这块土地并不紧密，但是却接受它的限制，将土地融入建筑内。由于有了一条不断收缩和拓宽的复杂通道，这个围合的空间成为一场探索之旅的高潮。在对泰俊公园的记忆中，自然和人文景观混合在一起，外部的灯光最终也成了能量的释放口。

Kimmel Eshkolot建筑师事务所设计的以色列赫茨尔山阵亡士兵纪念馆（168页）本身具有展览性质（巨大砖墙上刻着阵亡士兵们的名字），并将展览融入建筑中，以一系列复杂的空间参照定义了记忆和棺木。里伯斯金建筑事务所设计的加拿大国家大屠杀纪念碑（186页）的外墙规模庞大，变形的设计极具标志性，巨大的墙体赫然耸立在访客面前，展现了一种感觉位移的效果，也激活了一种反射机制。SANDWICH建筑事务所打造的科泰艺术展馆（120页）就采用了三维的围墙，这是一个盘旋在空中的空间围护结构。这样的建筑结构压缩了大部分内部空间，将黑暗作为反思和冥想的元素。用这样的形状来激活几乎无意识的回忆，正是阿尔多·罗西所采用的设计形式。

building houses the museum, enlarging the interior spaces to the outside, fading the space in a continuous landscape movement. The enclosure is the place of concentration, of reflection that amplifies remembrances because it centers people within a defined space, measuring and focusing on the people themselves: in the words of BCHO Architects – this distinctly Korean worldview is about embracement – an embracement of given conditions, of site and of surrounding – we can rediscover the meaning of architecture that seems to close: as in an embrace, instead, it protects and allows people to look to the outside, to the external reality.
In the Park Tae-joon Memorial Hall in Imnang Culture Park by BCHO Architects (p.70), the enclosure is an abstract space, which contains a piece of nature, as to simplify as much as possible disturbing elements, but defining a wall texture that produces a continuous reiteration of elements, such as a sort of mantra. It adapts to the topography of the land, favoring the internal views, but allowing the observation of external elements: it does not close to the territory, but rather embraces its limits, including them within the architecture. Thanks to a complex path, made of compressions and expansions, the enclosure becomes the culmination of a journey of discovery, where natural and artificial landscape mix together in the memory of Park Tae-joon and the outside light can finally unleash itself in its power. Mount Herzl Memorial Hall by Kimmel Eshkolot Architects (p.168) self-builds the exhibition element (the large brick wall with the names of fallen soldiers) and then encloses it within the building, defining the memory and the casket, in a complex series of spatial references; the National Holocaust Monument by Studio Libeskind (p.186) works on a marked deformation of the scale of the enclosure, contained within giant walls that loom over visitors, creating a sensorial displacement, which activates a mechanism of reflection; KOHTEI by SANDWICH (p.120) is a three-dimensional enclosure, a spatial envelope that hovers in the air. It compresses the interior space with force,

卡廷博物馆，波兰
Katyń Museum, Poland

安吉·雷迪博士纪念园，印度
Dr. Anji Reddy Memorial, India

通道，是激活记忆的最基础元素。步行，是梭罗和沃尔瑟作品中的最初动作，它可以帮助人进行持续的回想和记忆，它不但能刺激感觉，还能强有力地触发心理机制。

BBGK建筑事务所设计的卡廷博物馆（140页）就是对一块土地具化的项目。人们会在这里看到一条笔直且漫长的通道，它就像地面上的一道伤口般深邃；它的两侧被巨大粗糙的墙体包围，仅留下对天空的一瞥，使人处于极度专注的状态。Mindspace建筑事务所设计的印度安吉·雷迪博士纪念园（58页）则是以室内空间来定义室外空间，在人造景观中打造了一段由房间构成的漫长旅途，种种环境的混合，为访客们提供了一个放松、放空的环境。

作为布景要素的光的运用成为记忆概念中的基本成分。视觉刺激与尖锐阴影的对比，标记空间的照明效果，这些要素叠加在一起，通过精神状况的重塑，可帮助揭开人们的回忆。它们将对过去的研究定义为一种开放，这对记忆是十分重要的，它是每一座拥有纪念功能的建筑的基本要素。

在波兰Kwadrat建筑设计工作室设计的第二次世界大战博物馆（200页）中，光作为主角遍布长廊。这就像为富有深意的戏剧打造布景效果一样，利用光刺激访客的反应，同时也将其作为展览路线上参观房间的短暂停歇。

marking the darkness as an element of reflection, of meditation; it activates, with such defined shapes, an almost unconscious remembrance, as in the form of Aldo Rossi.

The path is a fundamental element in the incitement to memory: walking, which becomes the primary act in the writings of Thoreau and Walser, is the action that allows a continuous return of thoughts, remembrances, sensory stimulations, which strongly trigger mental mechanisms .

The Katyń Museum by BBGK Architekci (p.140) is a concretion of the land. It imposes a long straight path to people, a deep fissure in the ground, like a wound: it is enclosed by massive, rough walls, only focused to the sky, in an extreme act of concentration; Dr. Anji Reddy Memorial by Mindspace (p.58) defines exteriors in the form of interiors, a long journey made of rooms in the artificial landscape, in which a mix of environments creates a form of relaxation, a calm empathy for visitors.

The light used as a scenic element becomes a fundamental ingredient in a concept of memory. Visual stimulation, the contrast with sharp shadows, the luminous effect that marks the space, together they are able to recreate mental situations for the unveiling of remembrances. They define an openness to that research of the past, that is so important in Memento, and which is the basis of every architecture that hosts commemorative functions.

The Second World War Museum by Studio Architektoniczne Kwadrat (p.200) uses light as a protagonist in the long distribution corridor, making a scenic effect of profound drama, to stimulate a reflection in visitors, but also as a moment of pause along the several rooms of the exhibition itinerary.

于特岛赫恩胡塞特纪念馆和学习中心
Hegnhuset, Memorial and Learning Center in Utøya

Blakstad Haffner Architects

赫恩胡塞特或称为"安全之家",是一幢集学习、交流和纪念为一体的建筑。它是为了纪念2011年7月22日于特岛惨案而建造的。

赫恩胡塞特里包含着生与死的故事。有13人在咖啡馆建筑中失去了生命,但仍有19人幸存,建筑师Erlend Blakstad Haffner保留了这部分建筑。作为遗址的咖啡馆有助于那些受该事件影响的人及访客了解事件发生的经过及影响,也使建筑的记忆得到保存。

在老咖啡馆残存的部分上,建筑师加盖了第二层围护结构以作保护。新建筑的主体角度在设计时发生了扭转,与于特岛其他建筑的轴线相同。新建筑努力避开惨案中人们失去生命的那些地方。它代表并阐明了岛上的变化,一层新的历史,一页新的篇章。

支撑建筑屋顶的69根木柱子代表了2011年7月22日在惨案中去世的那些人,也正是它们使整个空间的意义更加明确。围绕这69根柱子,还有495根较小的柱子在室内空间外围形成安全防护围栏。495这个数字代表了在于特岛惨案中幸存下来的人,他们的余生也都难以忘怀这份沉痛的记忆。

访客可以漫步在外围栏和内支柱之间的走廊上,走廊的两端分别象征生者与死者。这条走廊表面上看是自由的,但实际却在强调一种被束缚的感觉。为了增强这种自由与束缚的不确定性,围栏的开口是随机放置的。同时,为了在建筑与景观之间创造安全感和流动性,建筑师在建筑中设置了五个入口。

这座建筑的材料突出强调了该事件留下的残酷记忆。由栅栏围起的房子没有进行抛光处理,而是由粗糙的、未经处理的木材和混凝土搭建而成的。屋顶最高有8m。

顶层是老咖啡馆建筑的残存部分,完整保留了穿过墙壁和家具的弹孔,敞开的窗户则是当初几个年轻人跳楼逃脱枪杀的地方。下面一层,设计师Atle Aas和Tor Einar Fagerland腾空了老咖啡馆下面的空间,并粗略地用混凝土覆盖了倾斜的地形,此处既是纪念馆,也是教育中心。老咖啡馆与新建筑之间的旧楼梯在"7·22"纪念区和环绕在外围的学习区域之间建立了视觉联系。

这幢建筑有效地完成了在展望未来的同时尊重过去的挑战。该设计将历史足迹和原始结构的基础要素结合到了一座崭新的建筑物之中。赫恩胡塞特融记忆与学习于一体的同时,也成为那桩惨案的永久见证。

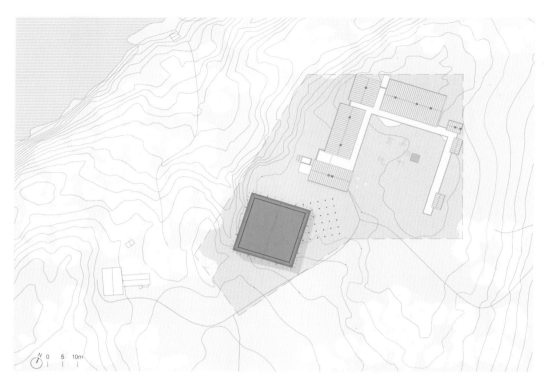

项目名称：Hegnhuset, Memorial and Learning Center, Utøya / 地点：Utøya, Hole, Norway / 事务所：Blakstad Haffner Architects
主要建筑师：Erlend Blakstad Haffner / 合作建筑师：Branko Belacevic, Vladimir Cvejic, Petar Stelkic, Ivana Barandovski, Bjørn Cappelen /室内设计：Atle Aas (exhibition), Siri Blakstad / 景观设计：IN'BY and Blakstad Haffner Architects /顾问：Tor Einar Fagerland, professor NTNU, Trondheim; Kurt Breitenstein; Alice Greenwald, director of the 9/11 Memorial Museum; James Young, professor, University of Massachusetts, Amherst; Edward Tabor Linenthal, editor, Journal of American History, professor, Indiana University; Clifford Chanin, Vice President for Education and Public Programs, 9/11 Memorial Museum; Jo Stein Moen, AUF veteran, author of Utøya book, communication communications manager at MARINTEK / 客户：Utøya AS/AUF / 总楼面面积：767m²
造价：2768 EUR/m² ex. VAT (2016) / 竣工时间：2016 / 摄影师：©Are Carlsen (courtesy of the architect)-p.106, p.107, p.110 p.112, p.117[upper], p.118[top], p.119; ©Espen Grønli (courtesy of the architect)-p.113, p.114~115, p.116; ©Hyun Yu-mi (C3)-p.104, p.108~109, p.111[right], p.117[lower], p.118[bottom-right]; courtesy of the architerct-p.118[middle, bottom-left]

The "Hegnhuset", or "Safeguard house" is a learning, communication and memorial house for the brutal incident that happened on Utøya, July 22nd, 2011.

The Hegnhuset contains a story of both survival and death. Architect Erlend Blakstad Haffner has preserved the part of the cafe building where 13 people lost their lives but also 19 people found refuge. The cafe building serves as a monument for those affected people and guests alike to visit and understand the impact of the events, to preserve an architectural memory.

Over the remains of the original cafe, a second building envelope was added as a protective cover. The new building body was laid in a distorted angle, in the same axis as other new buildings on Utøya. It is carefully placed to avoid the places where people lost their lives. It represents and clarifies change, a new historical layer and a new chapter in the island's history.

69 columns of wood that support the building's roof define the room and represent those who died on July 22nd. Around these 69 columns there are 495 smaller outer poles which create a safe, guarding fence around the interior. The number represents those who survived the tragedy on Utøya and who will carry thoughts and memories of this day for the rest of their lives.

Visitors can wander the corridor between the outside fence and the inner columns, symbolizing the dead and the living

repectively. The corridor emphasizes the feeling of being trapped, but apparently free. To enhance the feeling of uncertainty, the openings in the fence are placed randomly. To create a sense of security and flow between the building and the landscape there are five entrances to the building. The brutal incident is accentuated by the materials used in the building. The fenced house is not polished and consists of coarse untreated natural materials in wood and concrete. The ceiling height is 8 meters at the highest.

On the top floor are the remains of the old cafe building, with intact bullet holes through the walls and the furniture, and with open windows where several youths jumped out to escape the murderer. On the lower level, the underside of the cafe building has been hollowed out and the sloping terrain is roughly covered in concrete. The space now functions as both a memorial and an education center, designed by Atle Aas and Tor Einar Fagerland. The old staircase between the old cafe building and the new structure creates a visual link between the memories of July 22 and the learning zone around the house.

The architecture effectively meets the challenge of honoring the past while looking to the future. The design incorporates the footprint and foundational elements of the original structure into a new building. Hegnhuset embraces both remembrance and learning all the while bearing witness to the terrible events that happened at this very site.

残留建筑留作记忆 remains as memory

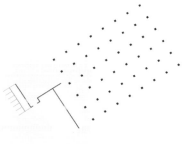

新场地 new ground

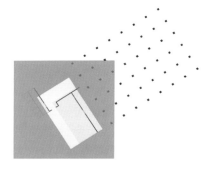

重建肌理 rebuilding context

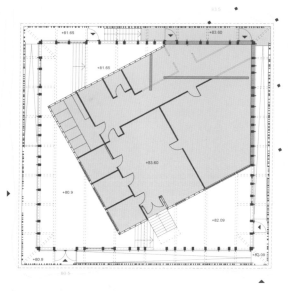

二层 first floor

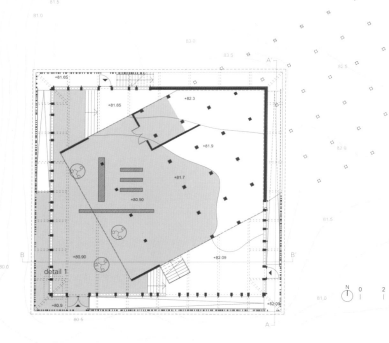

一层 ground floor

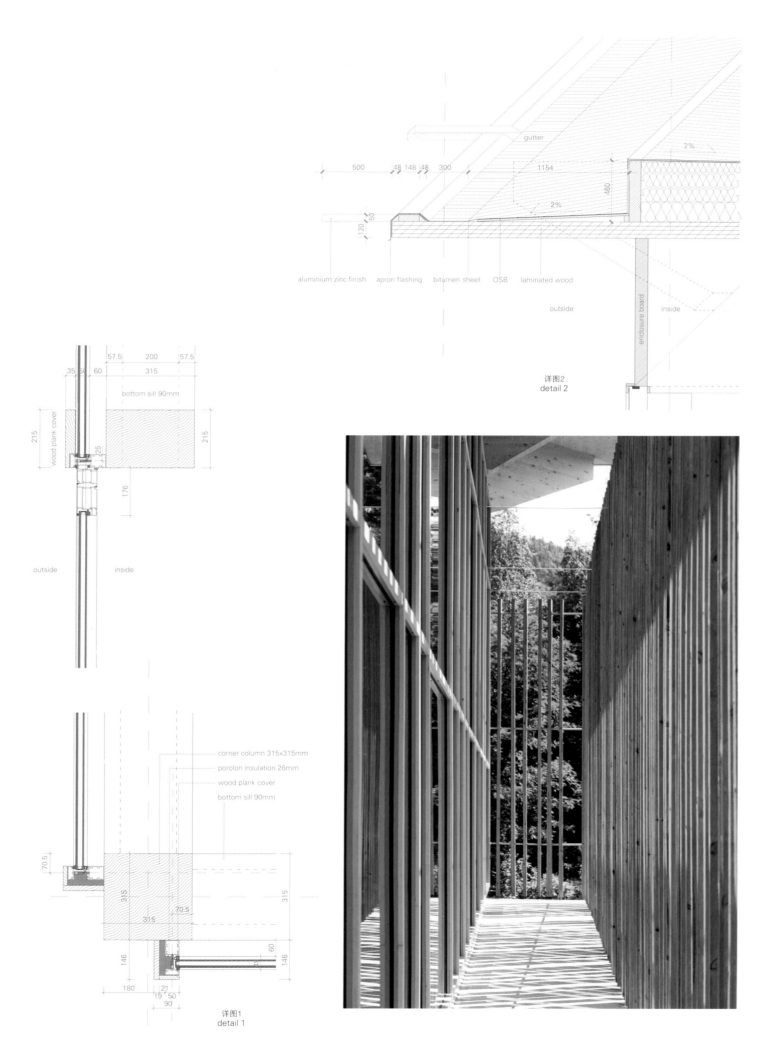

gutter

aluminium zinc finish　apron flashing　bitumen sheet　OSB　laminated wood

outside

enclosure board　inside

详图2
detail 2

bottom sill 90mm

wood plank cover

outside　inside

corner column 315x315mm
porolon insulation 26mm
wood plank cover
bottom sill 90mm

详图1
detail 1

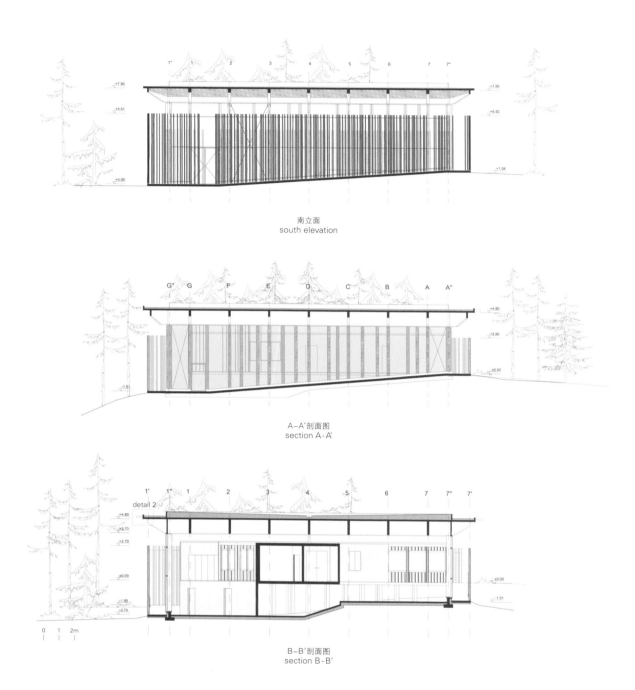

南立面
south elevation

A-A'剖面图
section A-A'

B-B'剖面图
section B-B'

科泰艺术展馆
KOHTEI

SANDWICH

　科泰艺术展馆坐落于日本广岛福田县新胜寺庙园区内的新胜禅宗博物馆与花园里。这座新胜寺庙是由一家造船公司建造的，用于慰藉那些在航海和造船事业中逝去的灵魂。访客们可以在这座综合建筑里体验一场难忘的禅宗之旅。通过感受其艺术装置，游客可以欣赏到园艺并体验冥想。科泰艺术展馆为游客们提供了一个深刻体会禅宗精神的机会。这一切只需要游客们通过一座人行桥既可抵达。科泰艺术展馆独特外形的灵感来源于寺庙建设的初始———座如船一般的建筑。这座建筑犹如在海浪上漂浮，周围可见群山，它仅由三种最基础的材料——木头、石头和水组成。

　展馆的主体建筑表面整个由日本杉木覆盖，它看起来像盘旋在整个景区之上，创造出底层的架空空间。拱腹内弯曲的表面向天空延展，很好地遮挡了阳光和风雨。屋顶的木板覆盖技术是千年之前在日本就存在的传统屋顶建造技术。这个木工制品使用了62万片沙原瓦，以使展馆的整体外观显得更为统一。

　在这样的空间中，在木屋顶梦幻般的轮廓的衬托下，景观朴实的物质感更得以凸显。周围的景色镶嵌了一些极富对比的元素，如古老的蕨类植物，使得游客可以体验到不断变化的风景。

　"走过这物质感满满的石头海洋，沿着一条缓缓上升的步行通道便可到达大楼的入口。进入室内，荡漾的海洋在黑暗的曙光中悄然呈现，"Kohei Nawa说道。

　游客们途经景观、花园和建筑，穿过走廊抵达一个狭小的入口后，就会渐渐地深入这座屋顶似船的建筑内部。在这里，在黑暗中，有一个代表着浩瀚海洋的装置。游客们可以一边体验冥想，一边观察反射在静静荡漾的水波上闪烁的灯光。这里有着微弱声响的黑暗反倒使游客们的视觉和听觉更加敏锐。体验持续25分钟，相当于和尚冥想练习的时间。在这里，每个人都会对冥想的时间和空间有不同的感觉。它的目的不是直接表达禅宗，而是帮助游客保留他们访问的记忆，并有机会认真思考禅宗的情感和哲学。

　科泰艺术展馆这个结构的外部、内部和架空空间给人一种置身于山间的环绕感觉，创造了一个将人们的身体体验和精神体验结合在一起的作品。它旨在为不可分割的建筑功能打造富有创意的表达，这是通过材料、纹理和建筑师积累的经验共同创造的。

KOHTEI is an art pavilion built in Shinshoji Zen Museum and Gardens within the campus of Tenshinzan Shinshoji temple in Fukuyama-city, Hiroshima, Japan. The temple was founded by the shipbuilding company to console the spirit of the dead in accidents at sea and industrial site. Visitors can expect a memorable Zen experience at the complex. KOHTEI offers visitors an opportunity to contemplate spirit of Zen by looking at its landscape and being subjected to a meditation like an experience through its art installation. Visitors make their approach through a connecting footbridge. KOHTEI's distinctive form was inspired by the roots of temple's establishment which led to create a building that resembles a ship. It's a building that floats on waves surrounded by mountains and is themed to work with three fundamental materials "wood", "stone" and "water".
The body of the pavilion is entirely covered with Sawara wood (Japanese cypress) that seems to hover above the landscape creating underneath a piloti space. The curved surface of the soffit expands to the sky warding off the sunlight and wind. The roof is covered with wood shingles us-

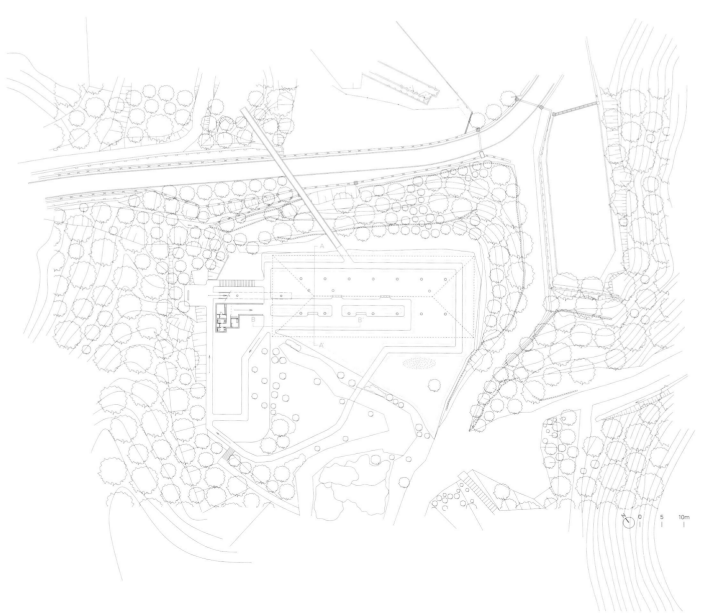

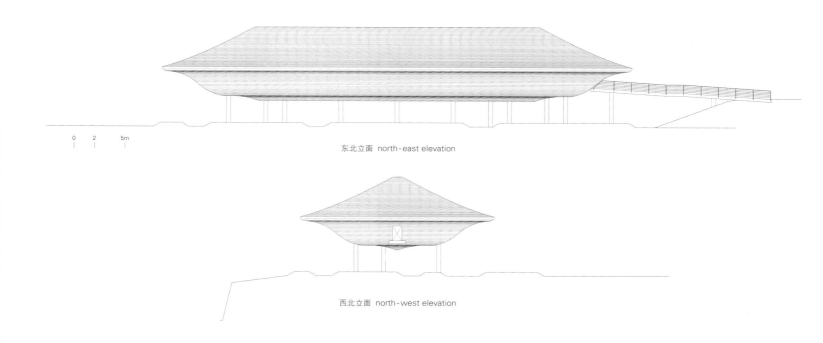

东北立面 north-east elevation

西北立面 north-west elevation

ing the Japanese traditional Kokera roofing technique. The woodwork is crafted by using 620,000 pieces of Sawara tiles in order to give a monolithic appearance to the pavilion. The experience of standing underneath such space enhances the stark materiality of the landscape against the airy contours of the wooden roof. Surrounding views are framed with contrasting elements including an ancient fern and visitors can experience ever changing scenery.

"Walking through the ocean of stones, full of materiality, one goes up the gently sloping walkway to reach the entrance of the building. Upon entering the interior, a quietly rippling ocean with glimmers unfolds in the darkness," says Kohei Nawa.

Visitors walk to the landscape, the garden and the building through the path, which gradually leads them into the interior of the vessel-like roof through a small entrance. Here, one finds an installation spreading in the darkness. The installation represents the immensity of the ocean and visitors can experience meditation while observing the shimmering lights reflected on the quietly rippling water waves. The darkness with the faint sound of the room, curiously sharpens visitors' vision and auditory senses. This installation lasts 25 minutes that is equivalent to the duration of monk's mediation practice period. Each individual will sense the meditative time and space differently. It was not intended to directly express Zen, but visitors retain the memories of their visit and have the opportunity to consider the sensibility and philosophy of Zen.

KOHTEI is a structure whose exterior, interior and underneath spaces reflect the enfolding experience of being in the mountains, creating a work that combines both physical and mental experiences. It aims to generate creative expressions of inseparably integrated architectural functions: the reality created by the materials and textures, and the experiences they engender.

项目名称：KOHTEI / 地点：Fukuyama-city, Hiroshima, Japan / 建筑师：SANDWICH Inc.–Kohei Nawa, Yoshitaka Lee, Yuichi Kodai / 项目经理：Office Ferrier / 结构工程公司：Asocoral Structural Engineering /机械工程：Yamada Machinary Office / 合作：NICCON, SUPER FACTORY / 规划：Shinshoji temple and Kohei Nawa with SANDWICH 屋面：Miyagawa Roof Company / 照明设计：Okayasu Lighting Design / 成本管理：Setouchi Holdings /承包商：Daiwa Construction Landscape / 监理：Sora Botanical Garden / 艺术装置：Kohei Nawa, WOW Inc. / 音响：Marihiko Hara / 平面设计：UMA/design farm / 庙宇场地面积：23ha / 场地面积：4700m² / 建筑面积：796m² / 尺寸：46m x 19m x 10m H 结构：steel / 屋顶：Kokerabuki(Sawara wood) / 拱腹：Wood Shingle (Sawara wood) / 设计时间：2013.4—2015.9 / 施工时间：2015.10—2016.9 摄影师：©Nobutada Omote (courtesy of the architect)

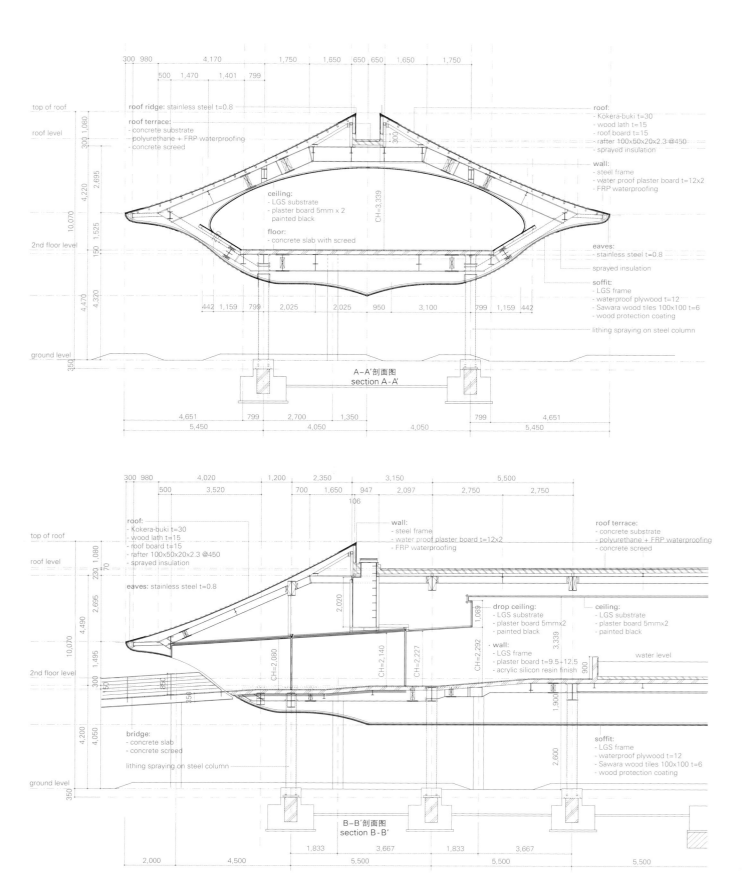

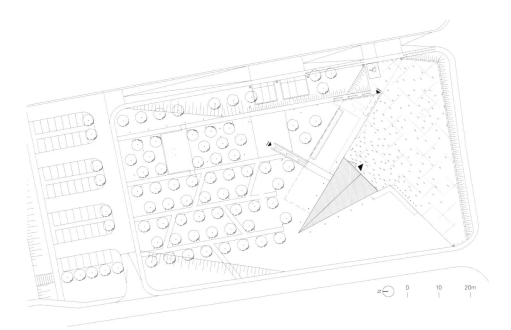

马科瓦的 Ulma 家庭博物馆
The Ulma Family Museum in Markowa

Nizio Design International

马科瓦的Ulma家庭博物馆是以第二次世界大战期间波兰人拯救犹太人为主题搭建，是波兰第一家纪念波兰人帮助犹太人的机构。

博物馆位于马科瓦的主干道旁，主要用以纪念1944年的"3·24"事件。第二次世界大战期间，纳粹宪兵在马科瓦枪杀了Wiktoria Ulma, Józef Ulma夫妇和他们的六个孩子以及藏在他们家中的犹太家庭。1995年，以色列Yad Vashem研究所授予Ulma一家"国际正义人士"的称号。

博物馆苦行般的建筑形式直切地面，展览区域就藏身其中。在博物馆的布局图上，我们不仅可以看到博物馆的结构形式，还能看到其他所有元素，如纹理和材料，这是对与博物馆信息相关内容的完好表达。钢筋混凝土建筑的立面覆盖着耐候钢板，打造出了一种铁锈般的外观，表现了时间的流逝。

嵌入地形之中的建筑形式与所使用的材料，使该建筑与周围融为一体。作为村庄和悠久历史的一部分，它让游客想起马科瓦战前的历史和生活。它不仅展现了大浩劫的那段时光，还揭示了一个道理：面对历史的厄运，自然存在的事物是不会发生变化的。

博物馆的立面一部分为玻璃，它是被简化成标志形式的大门。博物馆里可见的黄昏之光来自建筑中心。建筑的中心是一个玻璃长方体，象征着Józef Ulma和Wiktoria Ulma以及数千名冒着生命危险帮助犹太人的波兰人的家园。展品包括那时的家具、木工店、蜂箱、书籍、Józef Ulma的照相机和家庭文档。在这个空间里，还放映着一些反映这对夫妇及他们孩子日常生活的影像。Ulma一家作为代表之家，被安置在钢筋底座之上，外围的墙壁由安全玻璃制成，而安全玻璃上布满了薄薄的刻画，用来投影的墙上覆盖着反光膜，地板是用表面经过专门刷洗和老化的松木板制成的。

博物馆的参观路径围绕这个长方体和七个主题展开，并通过工具、文件、照片及手册和多媒体上展示的材料来讲述所发生的故事。展厅的中间有四个钢质立方体信息台，并设有触摸屏和座位。展览的所有内容都在讲述波兰人和犹太人在战争时期共同经历的悲惨命运。博物馆室内空间通过混凝土墙体分割，显得简洁、不朽且富有诗意。该博物馆最突出的是那道被照亮的垂直锐利缝隙（在展厅的后面），象征着一扇能够穿越无法理解的死亡区域的窄门。

院子旁边的纪念碑墙上设置了很多喷砂的花岗岩牌匾，上面写着那些拯救了犹太人的波兰人的名字。那些因拯救犹太人而丧生的波兰人的名字则突出地"嵌"在院子的地面中。博物馆入口处，发光的牌匾更为密集。进入院子里，那些牌匾就像河上的小船一样，排成了一队奇特的旅行灯，引导人们走向作为建筑立面标志的大门。

The Ulma Family Museum of Poles Saving Jews in World War II in Markowa is the first Poland's institution commemorating Poles who helped Jews.

The museum building occupies the site by the main road running through Markowa and commemorates the events of 24 March, 1944. During World War II in Markowa Nazi gendarmes shot Wiktoria and Józef Ulma, their six children and the Jewish families who had been hiding. In 1995, the Israeli Yad Vashem institute granted the Ulmas the title of the Righteous Among the Nations.

The museum's ascetic architectural form cuts into the ground and the exhibition hides inside. Within the museum's layout composition it is not only the form, but all the other elements, too, such as texture and material, that are to express the content related to the museum's message. The building of reinforced concrete has facades clad in weathering steel sheets which develop a rust-like appearance indicative of the passage of time. With the architectural form being recessed in the terrain and with the materials used, the building blends in with the surroundings and amalgamates with them. For becoming a part of the context of the village and the broader history it reminds visitors of the history and life of the pre-war Markowa. Not only does it refer to the time of the Shoah, but also reveals the unchanging nature of being, against the odds of fate and history. The partially glazed facade of the museum is a gate that is simplified to the form of a sign. Inside the museum there is twilight, illuminated by the glow of light coming from the heart of the building, which is a glass cuboid symbolising the home of Józef and Wiktoria Ulma, as well as the homes of thousands of Poles who risked their lives to help the Jews. The exhibits include the original furniture, a woodworking shop, a beehive, books, Józef Ulma's cameras and family documents. Within this space displayed are projec-

项目名称：The Ulma Family Museum in Markowa
地点：37-120 Markowa 1487, Poland
作者和主要建筑设计、室内设计、展览设计：Mirosław Nizio
项目成员：Mariusz Niemiec, Bartłomiej Terlikowski, Agnieszka Czmut, Witold Skarzyński, Andrzej Koper, Anna Derach, Natalia Romik, Katarzyna Okraszewska
景观建筑师：Studio Architektury Krajobrazu Viretum Agnieszka Michalska
表面面积：4,001m² / 建筑面积：626.80m²
可使用面积：512.76m² / 总面积：654.70m²
体积：3,118.20m³
施工：KOC-PROJEKT Zbigniew Koc
总承包商：Building–Skanska S.A.; Exhibition–MWE Sp. z o.o.
投资：Muzeum–Zamek w Łańcucie, ul. Zamkowa 1, 37-100 Łańcut
竣工时间：2016
摄影师：©Lech Kwartowicz (courtesy of the architect)

tions that bring back the scenes from everyday life of the married couple and their children. The symbolic home of the Ulmas is perched on a steel substructure, and the walls are finished with safety glass covered with engravings on film substrate. The wall on which projections are displayed is covered with anti-reflective film. The floor is made of pine boards with brushed and aged surfaces.

The viewing path of the museum leads around the cuboid and across the 7 thematic sections, where the story is told through artefacts, documents, photographs, and materials

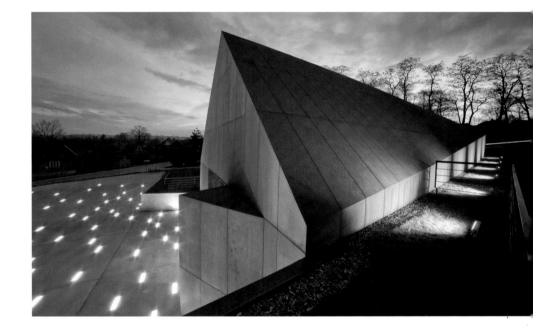

presented at manual and multimedia stands. In the middle of the exhibition room are 4 infoboxes in the form of steel cubes with touch screens and seats. All elements of the exhibition are arranged so as to tell the story of the shared past of Poles and Jews in the context of the tragic time of war. The interior of the museum is kept in simple and monumental poetics of concrete walls. Its culmination – at the back of the exhibition room – is the illuminated vertical and sharp gap which symbolises the narrow gate that leads through the incomprehensible area of death.

On the monumental wall adjacent to the yard placed are sandblasted granite plaques featuring the names of the Poles who saved Jews. Then, "embedded" in the very plane of the yard are highlighted plaques with the names of those who lost their lives for saving Jews. The density of the illuminated plaques increases towards the entrance to the museum. In the yard, like boats on a river, they form a peculiar procession of travelling lights that approach the threshold of the gate that is symbolised by the house elevation.

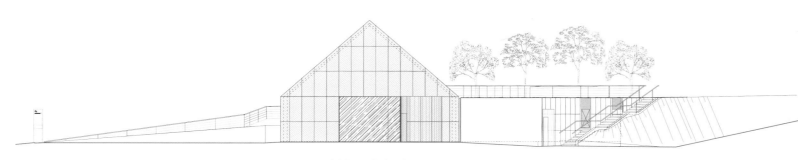

南立面 south elevation

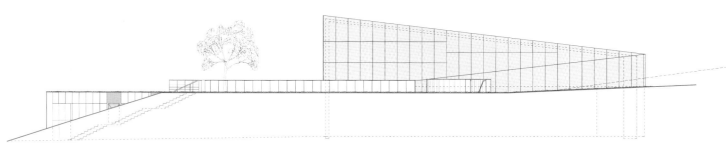

东立面 east elevation

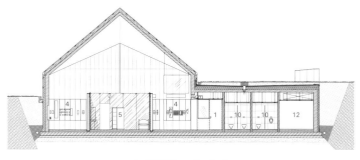

A-A'剖面图 section A-A'

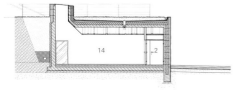

B-B'剖面图 section B-B'

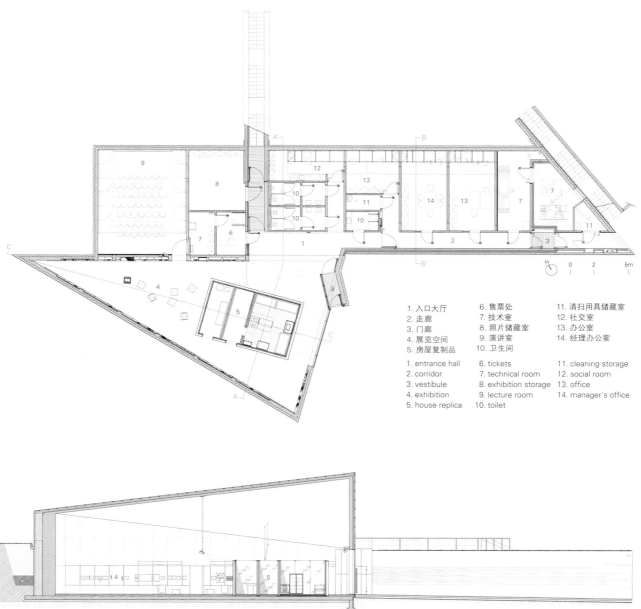

1. 入口大厅	6. 售票处	11. 清扫用具储藏室
2. 走廊	7. 技术室	12. 社交室
3. 门廊	8. 照片储藏室	13. 办公室
4. 展览空间	9. 演讲室	14. 经理办公室
5. 房屋复制品	10. 卫生间	
1. entrance hall	6. tickets	11. cleaning storage
2. corridor	7. technical room	12. social room
3. vestibule	8. exhibition storage	13. office
4. exhibition	9. lecture room	14. manager's office
5. house replica	10. toilet	

C-C'剖面图 section C-C'

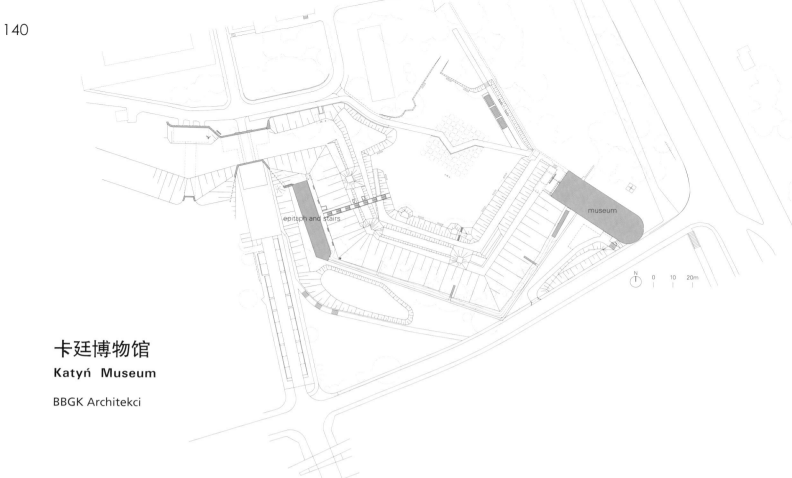

卡廷博物馆
Katyń Museum

BBGK Architekci

本案博物馆是利用先前已有的建筑达到设计目的的杰出典范。在市中心的一个军事堡垒中，能隐隐约约看到远处的一片森林，这片森林见证了恐怖的卡廷大屠杀。博物馆里的展品以人们在书中或纪录片中常见的形式，讲述着卡廷的故事。

卡廷博物馆记载着第二次世界大战期间在卡廷所发生的大屠杀事件，以此纪念被杀害的22 000名波兰战俘。

该博物馆位于19世纪的华沙城堡南部，包含三座历史建筑。整个建筑体设计成公园形式，在主广场的中心建有象征性的卡廷森林。那里种了100棵树，用以呈现这场可怕战争罪行的真相，因为这些过去的真相在那森林里隐藏了50多年之久。主展览设在两层的卡波尼式的历史堡垒中。第一层主要介绍了有关卡廷大屠杀的信息，游客不仅可以在那里了解历史的真相，还能看到从卡廷森林的乱葬坑中发现的展品。展览的第二层是出于人道主义专门为受害者家属所建的用来沉思的场所。

这里使用的材料为砖、石膏和染色混凝土，而不是那种纪念碑常用的不朽材料，因为只有这样才能赋予这个地方一种谦逊和沉稳的气氛。从天空俯瞰，其深挖的形式极富戏剧性，与人为留下的印记和刻在墙上的名字形成了鲜明的对比。

博物馆的出口由Jerzy Kalina设计，它是一条由黑色混凝土建造成的20m长的"死亡隧道"。这条黑暗的走廊通往名为"失踪的人"的小巷。"失踪"是因为巷子里布满空空的台座，上面仅刻有死者的职业："警官、医生、律师、建筑师……"而非姓名。穿过小巷，便是第三座建筑。这里有弧形的大炮架，有镶着琉璃的拱廊，上面陈列着15个写着21 768名被杀害的军官名字的牌匾。建筑师们用染了色的混凝土作为他们展现建筑的表达手段。受害人的部分信件和其他个人物品被刻印在混凝土墙壁中，将展览一直延伸到建筑之外，建筑风格表达在这里展现得格外强烈。12m高的墙体之间的空隙将参观城堡之路引向两个不同的方向：下边，是刻着受害者名字的牌匾；上边，是天空和阳光。设在树林中的一个橡木十字架结束了这段戏剧性的卡廷故事。

建筑的初衷不仅在于建造墓地和纪念场所，也在于打造避难场所。卡廷博物馆带给我们人去楼空的感觉。这座建筑就像你步入森林里时会看到的墓地标志一样（这正是阿道夫·卢斯对目的纯粹的建筑的定义），令人印象深刻，清晰明确，尽管它并非位于原本所在的地方。它时刻提醒着我们，纪念性建筑的核心在于——转移、拆除和重现。

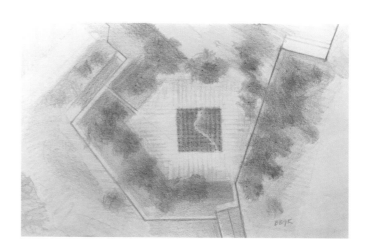

The museum constitutes an outstanding example of how pre-existing architecture can be employed to serve the designed purpose. The faraway forests which had witnessed the horrible Katyń massacre were somehow made present in a military fortification in the center of the city. The exhibits tell the story of Katyń in a manner one could find in a book or a documentary.

The Katyń Museum presents the tragic events of the Katyń massacre that took place during World War II and commemorates 22,000 Polish Prisoners of War murdered.

The museum is located in the southern part of a 19th-century fortress – the Warsaw Citadel – and includes three historical buildings. The whole complex was designed as a park, with a symbolic Katyń forest in the center of the main square. The 100 trees planted there refer to the truth about the dreadful war crime, which used to be concealed in the woods for more than 50 years. The main exposition is arranged on two levels of the Caponier – a historic fortification structure. The ground level contains information about the Katyń massacre, where visitors can learn about historical facts and see exhibits found in the mass graves in Katyń forest. The first level of the exhibition is devoted to personal tragedies of the victims' families, constituting a place for contemplation.

The use of brick, plaster, and stained concrete, rather than the more monumental materials that are usual for such memorials, gives the place a humble and contemplative atmosphere. The deep cut with its view of the sky is truly dramatic, and stands in stark contrast to the delicacy of the imprints left by the personal effects and the ephemeral quality of the names incised in the wall.

The exit of the museum turns into the Death Tunnel – a 20-meter-long passage constructed from black concrete, designed by Jerzy Kalina. This dark corridor opens towards the Alley of the Missing Ones. "Missing" because the alley is filled with empty pedestals, on which only the professions of the deceased are engraved: "police officer, doctor, lawyer, architect…" The path leads further to the third building – the arcaded cannon stand with glazed arcades, displaying 15 plaques with the names of the 21,768 murdered officers. The architects used stained concrete, turning it into a means of architectural expression. Parts of letters and other personal belongings of the victims are imprinted on the concrete, continuing the exhibition outside the buildings. The architectural expression here is especially strong. The gap between 12-meter-high walls dividing the Citadel leads in two directions: down-towards the plaques with the victims' names, and up-towards the sky and light. An oaken

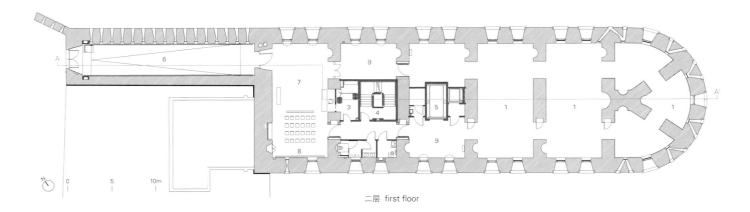

二层 first floor

1. 展览空间　　　1. exhibition space
2. 出口　　　　　2. exit
3. 外套传送带　　3. coat conveyor
4. 楼梯　　　　　4. staircase
5. 电梯　　　　　5. lift
6. 入口　　　　　6. entrance
7. 入口空间　　　7. entrance space
8. 展示屏　　　　8. display screen
9. 展室　　　　　9. exhibition room
10. 办公空间　　 10. office space
11. 储藏室　　　 11. storage
12. 厨房　　　　 12. kitchen
13. 绿色露台　　 13. green terrace
14. 机械室　　　 14. mechanical room
15. 电气室　　　 15. electrical room
16. 聚集空间　　 16. aggregates
17. 供暖室　　　 17. heating room

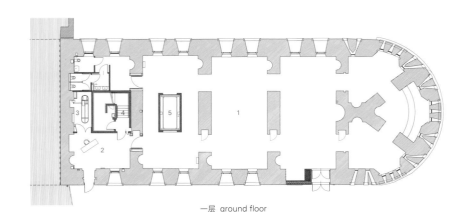

一层 ground floor

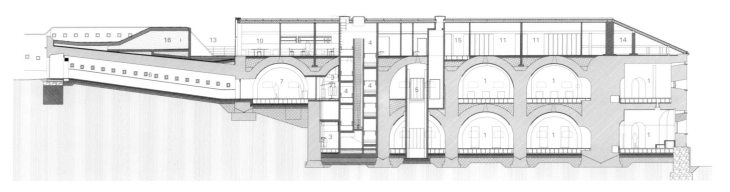

A-A'剖面图 section A-A'

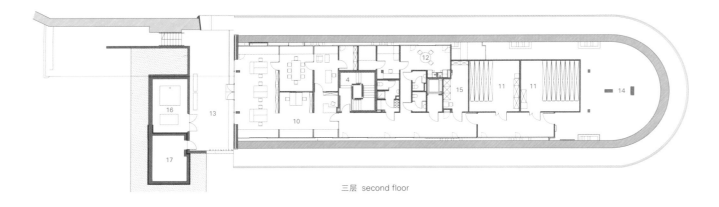

三层 second floor

项目名称：Katyń Museum
地点：Jana Jeziorańskiego, Warsaw, Poland
建筑师：BBGK Architekci
项目团队：BBGK Architekci–Jan Belina Brzozowski, Konrad Grabowiecki; Plasma Project s.c.–Jerzy Kalina (original installations of plastic narration), Justyna Derwisz, Adam Kozak; Maksa–Krzysztof Lang
合作者：Joanna Orłowska, Marek Sobol, Emilia Sobańska, Łukasz Węcławski, Agnieszka Grzywacz, Ewelina Wysokińska, Jacek Kretkiewicz, Tomasz Pluciński, Maciek Rąbek, Marcin Szulc, Barbara Trojanowska, Jolanta Fabiszewska / Structural engineer: BBK Piotr Szczepański
电气照明工程师：Candela BI / 景观建筑师：Anna Kalina, PASA Design, Małgorzata Ogonowska / 总承包商：PNB Południe
客户：Muzeum Wojska Polskiego
用地面积：27,000m² / 建筑面积：1,808m² / 总楼面面积：23,966m²
设计时间：2010~2013 / 施工时间：2013~2015 / 竣工时间：2016
摄影师：©Juliusz Sokolowski (courtesy of the architect)

cross placed among the trees concludes the dramatic story of Katyń.

The origins of architecture lie as much in tombs and memorials as they do in a shelter, and the Katyń Museum takes us back to that sense of the fixing in one place of lives that were lived, but are now gone. As a grave marker you come across in the forest, which was Adolf Loos' definition of pure architecture, it is effective and clear, despite its remove from the actual location. It reminds us of the alienating, removing, and making present that are at the core of monumental architecture.

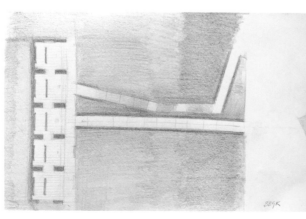

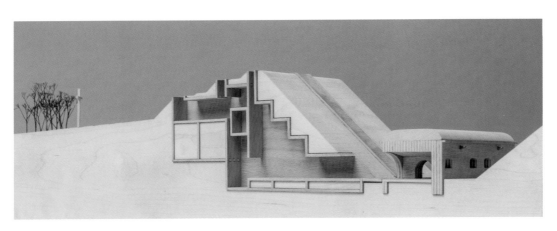

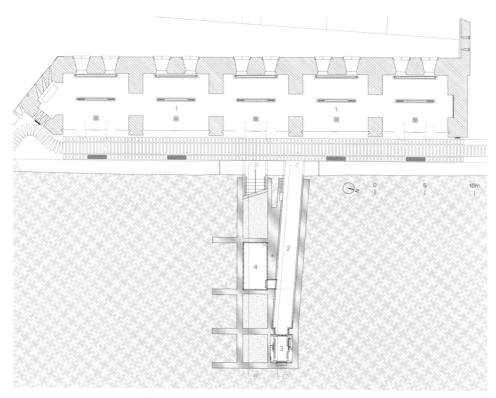

1.展览空间 2.隧道 3.电梯 4.设备间　1. exhibition 2. tunnel 3. lift 4. equipment
碑文与楼梯平面图 epitaph and stairs plan

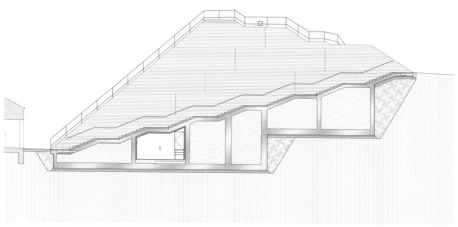

1.设备间　1. equipment
B-B'剖面图 section B-B'

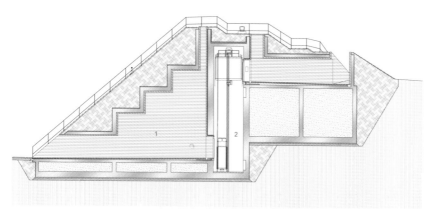

1.隧道 2.电梯　1. tunnel 2. lift
C-C'剖面图 section C-C'

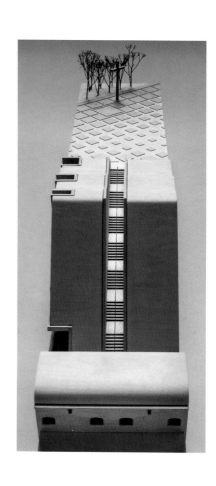

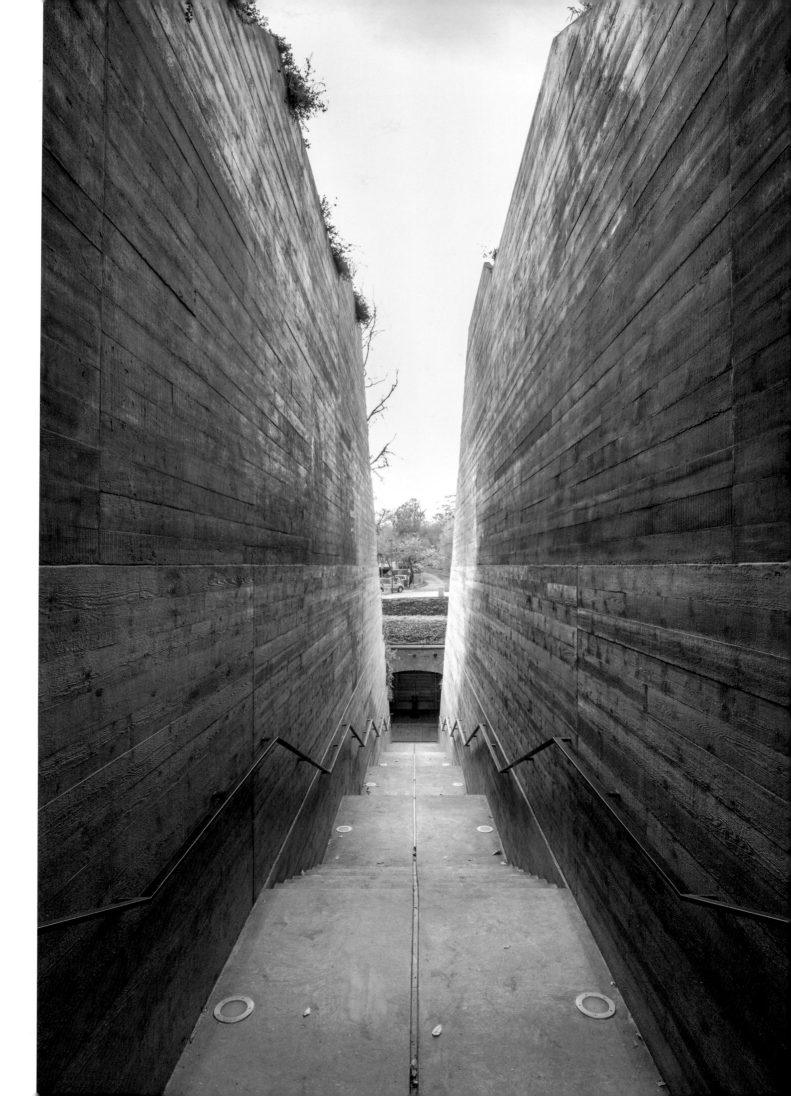

大地的见证——侵华日军第七三一部队罪证陈列馆
The Witness of Land
– Memorial Hall of Crime Evidences by Unit 731
华南理工大学建筑设计与研究所 Architectural Design & Research Institute of SCUT South China University of Technology

侵华日军第七三一部队犯下了诸如生化战和人体实验等罪行。一九四五年八月，日本投降前夕，侵华日军第七三一部队在逃亡前轰炸并摧毁了大部分建筑物，形成了现在这片废墟的整体格局。

由于快速的城镇化，这片遗迹周围已经变成中心住宅区。因此，如何界定这个博物馆和这座城市的关系成为亟待解决的问题。建筑师在场地东侧设计了一座街道公园，通过降低博物馆与周围空间形态之间的反差使整体环境形成更好的过渡。这座街道公园成为这片遗址和所在城市之间的过渡带。街道围墙能够过滤一部分外界噪声，但由于遗址入口不宜搭建围墙，建筑师设计了一个下沉广场以形成一个相对安静的物理空间。遗址西侧，一条横贯南北的铁路仍在运行。为了在不破坏原始环境的同时使新建筑与这片遗迹建立更紧密的联系，建筑师复原了之前的道路网格，以混凝土和夯土代替用土砖和铁丝网建成的围墙。除了这片遗迹之外，建筑师还用灰色沙砾铺路增加这片区域的衔接性。设计原则在于保存遗迹同时修复特别部分以重建那段时期的历史场景。

为降低陈列馆的高度和体量感，其主要陈列空间沉入基地，入口下沉以减轻对城市的影响。这座建筑静谧而有力地介入场地之中，为这座城市营造了一种别样的氛围，它在以平静的态度表达人类历史。

"黑盒子"这个概念基本上是对侵华日军第七三一部队所犯罪行的揭露。东京审判期间事件还未曝光，日军暴行在之后才逐渐为人所知。这一过程就像飞机失事之后寻找黑盒子和还原事故经过。所以黑盒子代表着一旦打开就会揭露所有真相的真相记录器。

这座黑色建筑有着标志性的深灰色倾斜屋顶，这让它看起来像是覆盖砾石的地面的延伸。这个"倾斜"不仅展示了建筑尺寸，还体现出一种设计态度。从办公大楼前面的广场向东望去，陈列馆看似地面的一个轻微波动，比起人造几何形体更像是一种天然大地景观。裂缝使整座黑盒子建筑在形式、颜色和材料上形成了微妙差别，这反过来也能帮助这座建筑融入环境之中。

黑盒子坍塌，下沉并在场地上破碎，留下一道永不磨灭的伤痕，仿佛一把锋利的手术刀把地面切割开来。这一系列简单的设计行为建立了建筑内外的联系，同时创造了一个可交流的对话空间。这样，罪证陈列馆作为一个客观容器，能够使游客感知理解历史事件和这座建筑的深意。

The Unit 731 of the Japanese Imperial Army that invaded China committed crimes such as chemical and biological warfare and human experimentation. In August of 1945, on the eve of Japanese's surrender, the 731 Unit bombed and destroyed a large portion of buildings before fleeing, forming the overall configuration of the current ruins.
Due to rapid urbanization, the surrounding area around the ruins has become a central region for houses. Hence, defining the relationship between the museum and the city was the major issue needed to be solved. A street park was designed on the east side of the site, forming a better transition of the environment by reducing the contrast between the museum and surrounding spatial condition. The park serves as a transition between the site and the city. The

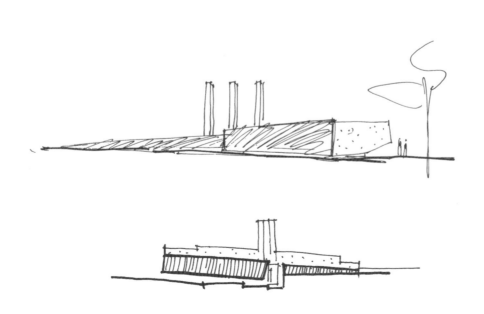
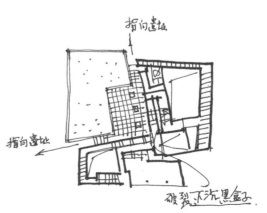

remained ruins
barracks' boundary wall

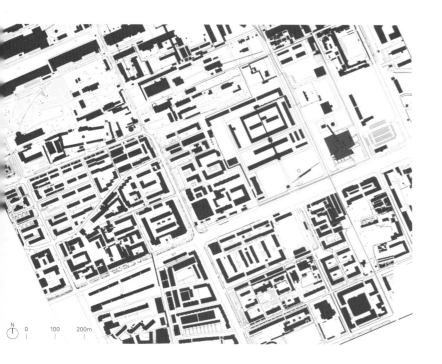

N 0 100 200m

项目名称：The Witness of Land – Memorial Hall of Crime Evidences by 731 / 地点：Pingfang District, Harbin, Heilongjiang Province, China
建筑师：Architectural Design & Research Institute of South China University of Technology / 主要创意建筑师：He Jingtang, Ni Yang, He Chili, He Xiaoxin, Liu Tao
结构工程师、机电管道顾问、工料测量师、其他顾问：Architectural Design & Research Institute of South China University of Technology
照明工程：GH Architectural Lighting Design Ltd. / 总承包商：Heilongjiang Construction and Installation Group Co.,Ltd
客户：Memerial Hall of Crime Evidences by Unit 731 of the Japanese Imperial Army / 总楼面面积：9,997m² / 造价：12 billion RMB
设计时间：2014.4—2015.1 / 施工时间：2014.9—2015.8 / 摄影师：©Yao Li (courtesy of the architect)

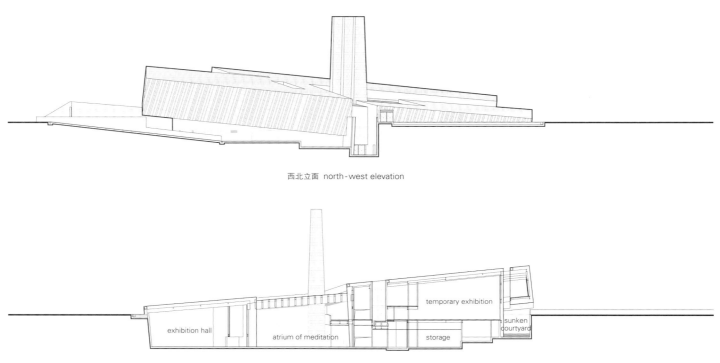

西北立面 north-west elevation

A-A'剖面图 section A-A'

enclosing wall along the street is also able to partially filter the noisy outside. Since it is not proper to build an enclosing wall at the entrance to the site, the architect designed a sunken plaza to be a relatively tranquil physical space.
On the west side of the site, a railway that cuts through the site from south to north is still in service. In order to establish a closer connection with the ruins beyond the railway, at the same time not to damage the original context, the architect reinstated the previous road network, then replaced the enclosing wall made of adobes and wire net with concrete and tamped earth. Except for the ruins, grey gravel was used for pavement to give continuity in this area. The design principle is to preserve the ruin as it is while restoring particular sections to recreate the historical scenes of the period.
In order to reduce the Memorial Hall's height and the sense of existence, the main display spaces are sunk into the ground with submerged entrance to filter out the city. The building intervenes with the site quietly yet powerfully, creating a new atmosphere with the city. It is expressing the history of humanity with a quiet attitude.
The concept of the "Black Box" is basically an unearthing of the crimes committed by Unit 731. The incident was not exposed during the Tokyo Trial, and the atrocities came to people's knowledge gradually afterwards. The process is like finding the black box and restoring the accident after an air crash. So the Black Box represents a truth-recording container, which will disclose all truth once being opened.
The dark volume features a slanted form with a charcoal-grey roof that appears as an extension of the gravel-covered terrain from which it emerges. The "tilt" shows not only the scale but also the attitude of design.
Viewing from the square in front of the office building into the east, the exhibition hall seems like a slight fluctuation in the terrain, more of a natural earth-scape than an artificial geometry. The crack creates subtle variations in form, color and material of the Black Box, which in turn help merge the building into the environment.
The Black Box collapses, sinks, and falls apart on the site, leaving an eternal "scar" as if the land was incised by a sharp scalpel. These series of simple acts establish interior-exterior building connection while creating a communicative and dialogic space. Here the building serves as an objective container for visitors to perceive and interpret the historical event and the implication of the architecture.

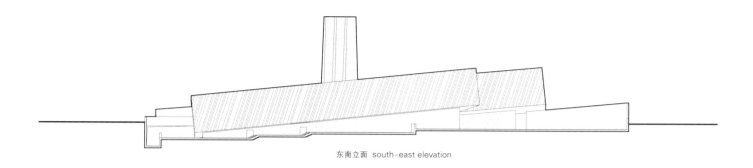

东南立面 south-east elevation

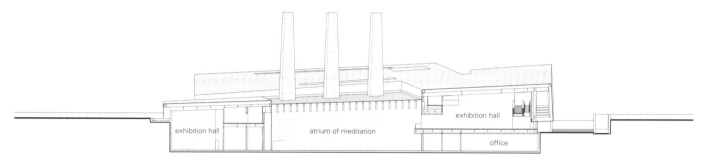

B-B'剖面图 section B-B'

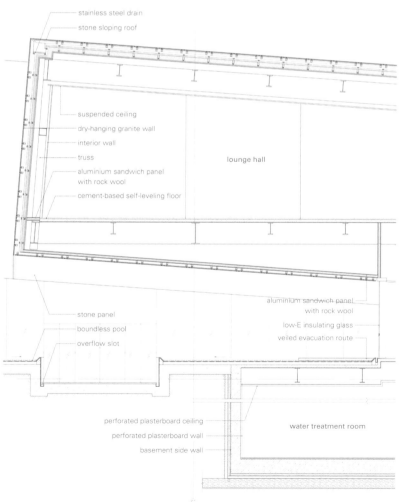

a-a'详图 detail a-a'

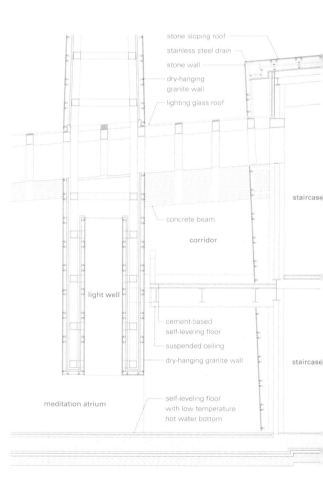

b-b'详图 detail b-b'

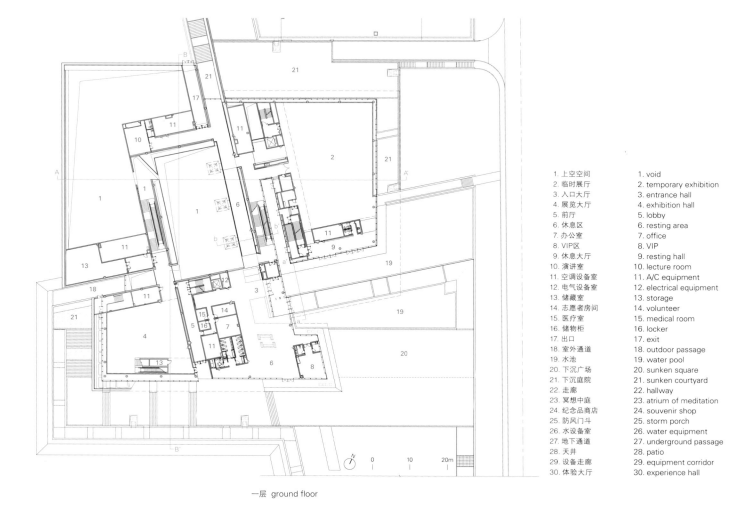

1. 上空空间	1. void
2. 临时展厅	2. temporary exhibition
3. 入口大厅	3. entrance hall
4. 展览大厅	4. exhibition hall
5. 前厅	5. lobby
6. 休息区	6. resting area
7. 办公室	7. office
8. VIP区	8. VIP
9. 休息大厅	9. resting hall
10. 演讲室	10. lecture room
11. 空调设备室	11. A/C equipment
12. 电气设备室	12. electrical equipment
13. 储藏室	13. storage
14. 志愿者房间	14. volunteer
15. 医疗室	15. medical room
16. 储物柜	16. locker
17. 出口	17. exit
18. 室外通道	18. outdoor passage
19. 水池	19. water pool
20. 下沉广场	20. sunken square
21. 下沉庭院	21. sunken courtyard
22. 走廊	22. hallway
23. 冥想中庭	23. atrium of meditation
24. 纪念品商店	24. souvenir shop
25. 防风门斗	25. storm porch
26. 水设备室	26. water equipment
27. 地下通道	27. underground passage
28. 天井	28. patio
29. 设备走廊	29. equipment corridor
30. 体验大厅	30. experience hall

一层 ground floor

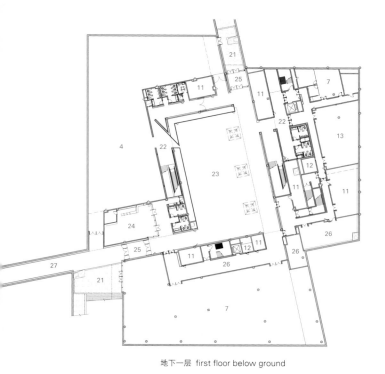

地下一层 first floor below ground

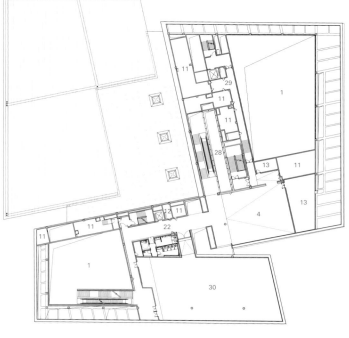

二层 first floor

赫茨尔山阵亡士兵纪念馆
Mount Herzl Memorial Hall

Kimmel Eshkolot Architects

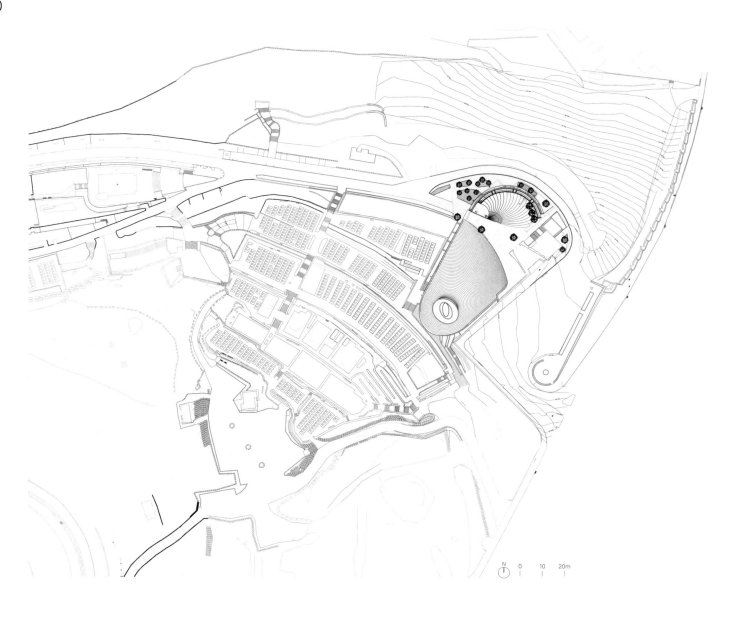

项目名称：Mount Herzl Memorial Hall
地点：Jerusalem, Israel
建筑师：Kimmel Eshkolot Architects
建筑师负责人：Etan Kimmel, Limor Amrani
合作者：Kalush Chechick Architects
结构工程师：Haim & Yehiel Steinberg Structural Engineering
施工经理：E.D. Rahat Engineering Coordination and Management Ltd.-Eliezer Rahat, Daniel Rahat
设计阶段经理和协调人：Eran Garber E.S.L Engineers
承包商：Green Construction Ltd-Nadav Rubin _ Project manager, Eran Rosenberg _ Engineer
光筒优化：R/O/B Technologies, IDF Merkava and Armored Vehicles Directorate
照明设计：Amir Brenner Lighting Design
3D模型制作：XENOM
客户：Ministry of Defence - families and Commemoration Department
用地面积：3,500m²
建筑面积：5,000m²
结构：concrete, steel columns
材料：concrete, steel, Jerusalem stone
设计时间：2010
施工时间：2015—2018
摄影师：©Amit Geron (courtesy of the architect) (except as noted)

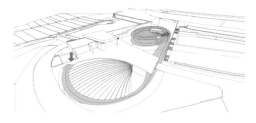

交通路线 circulation

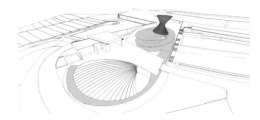

光筒 the funnel

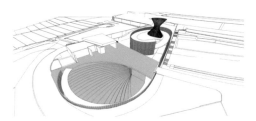

姓名墙 the wall of names

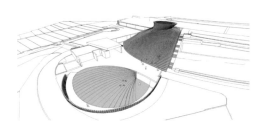

石材屋顶再塑原有地形
the stone roof recreating the original topography

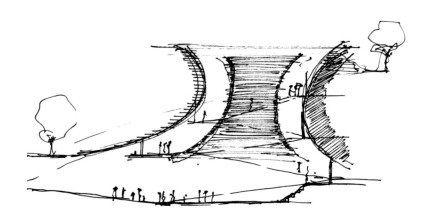

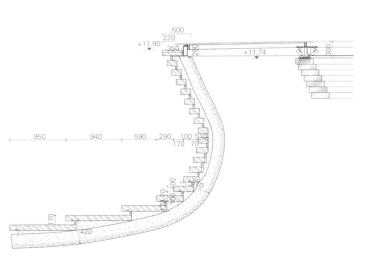

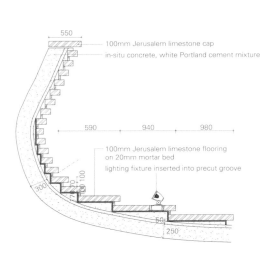

详图1 detail 1

赫茨尔山阵亡士兵纪念馆竣工于2017年9月，挖掘自耶路撒冷山山体，形成一个可供个人或集体进行纪念活动的私密空间。纪念馆上方，山体被重建成为一个由砖块砌成的好似波动着的漏斗形，开口朝向天空，筒内充满自然光。

除了纪念馆的空间之外，设计理念的第二个关键元素是一面长达250m连续不断的"姓名墙"，它包裹着中心极具雕塑感的砖结构。这面墙沿着螺旋坡道盘旋而上至纪念堂。墙体由23 000块砖砌成，每一块砖都刻着一位阵亡士兵的名字和死亡日期。每逢士兵的忌日，对应砖块上的蜡烛就会点亮。为了给纪念馆创造一个静谧的空间，纪念馆的内部设计就成了第一项工程。下挖山体让充足日光进入是首要任务，重建地形位居其次。这项工程志在外观雄伟壮丽，又与毗邻公墓在感观上相呼应。建筑内部，从顶部光眼进入的自然光经过光筒的巧妙过滤后进入空间。漏斗形光筒由8000块砖建成，这些砖由40cm×10cm的定制挤塑铝型材切割而成。通过这些"石头花边"的光线在大厅内摇曳闪烁，并照射在姓名墙上。

虽然纪念馆紧邻如今非常繁忙的耶路撒冷街道，但其环境宁静而神圣，主要的纪念空间显得孤僻而寂静。参观者可以直接穿过它去往赫茨尔山烈士公墓，或者走下台阶沿着砖墙到达纪念堂入口，砖墙内侧就是姓名墙。

纪念馆被规划为一座纪念性建筑，因此，设计的重点是创造一种几乎不需要任何机械系统，而且几乎完全可持续的非建筑。建筑中没有空调或者电力通风系统。自然气流创造出极佳的温度条件，利用漏斗形状通过间隔的石板把热气从屋顶上端排出，从而创造自然气流为建筑通风。白天照明完全无须用电，自然光从光眼进入，经过光筒的巧妙过滤，使整个空间充满了舒适的光线。纪念馆由山体挖掘而成，拥有最佳的热学条件。该结构的蓄热能力强，与地表一起，能使建筑保持稳定的温度。室外的地面用浅色的耶路撒冷石覆盖，这种石头可保护建筑不受辐射。使用诸如耶路撒冷石之类的当地材料是设计过程中的一个关键元素，不仅为了节省开支，还考虑到可持续性，旨在让建筑与周围的耶路撒冷城融为一体。

Mount Herzl Memorial Hall, finished in September 2017, was excavated in the Jerusalem mountain to form an intimate space for a personal and collective experience of commemoration. Above the hall, the mountain was reconstructed with an undulating funnel shaped formation of bricks which opens the hall to the sky, flooding the void with natural light.

The second key element to design concept, apart from the space of the hall, was a 250-meter-long continuous "Wall of Names" that wraps around the central sculptural brick structure. Following a spiral ramp up the memorial, the wall is built of 23,000 bricks, each engraved with the name of a fallen soldier, the date the soldier was killed and a candle to

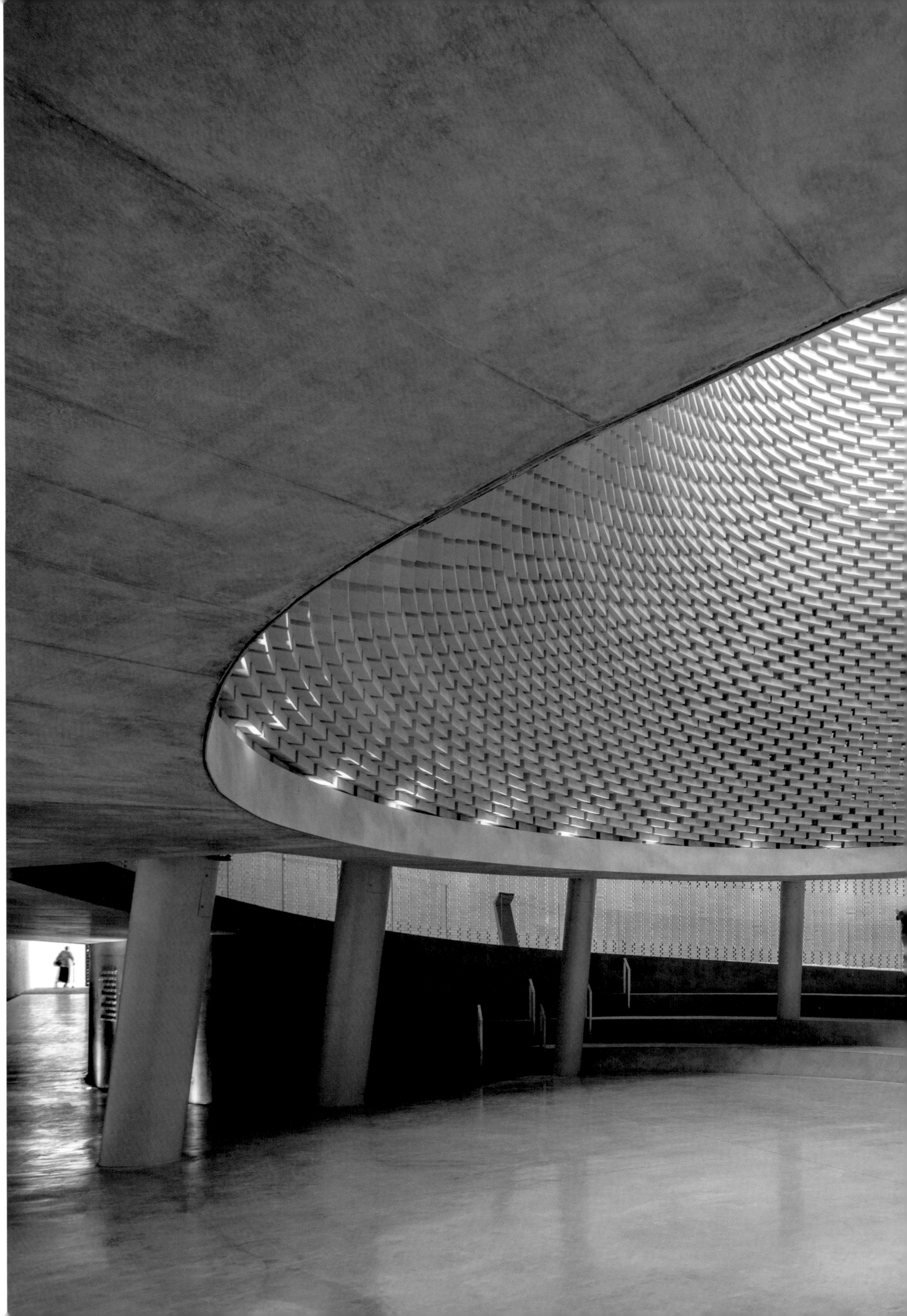

be lit on the anniversary of the soldier's death.

With the intention of creating a serene space for the hall, the design of the memorial is firstly a project of an interior. Excavating the mountain to allow for abundance of daylight in the hall was the primary focus, with the reconstruction of the terrain being secondary. It is intended to be unimposing from the exterior, and to echo the texture of the adjacent cemetery. In the interior, the light that enters through the oculus is subtly filtered through the funnel. The funnel is constructed of 8,000 bricks which were cut from a custom made extruded aluminium profile of 40x10 cm. Through this "stone lace", light flickers in the parameter of the hall, on the Wall of Names.

Set in a calm biblical scenery, but adjacent to the nowadays busy streets of Jerusalem, the main commemoration space is isolated and quiet. Visitors can walk directly through it towards Mount Herzl Cemetery, or walk down a series of stairs to the entrance of the hall, along a brick wall which in the interior becomes the Wall of Names.

The Memorial was planned as a monument, and as such, the design focused on creating a non-building that can function almost without any mechanical systems and be close to entirely sustainable. There are no air-conditioning or electrical ventilation systems. Natural air flow creates excellent temperature conditions using the funnel shape to expel hot air out of the upper end of the roof through spaced stone slabs, thus creating air flow that ventilates the place. There is zero use of electricity for daylight. Natural light enters through the oculus and is subtly filtered through the funnel of light, flooding the space with pleasant light. Excavated in the mountain, the Memorial Hall obtains optimal thermal conditions. The structure's thermal mass, integrated within the earth, keeps a steady temperature. The exterior topography was clad with light-coloured Jerusalem stone, which protects the building from radiation. Using local materials such as the Jerusalem-stone was a key element in the design process both for budget constraints and for sustainability reasons, aiming to integrate the building with the surrounding city of Jerusalem.

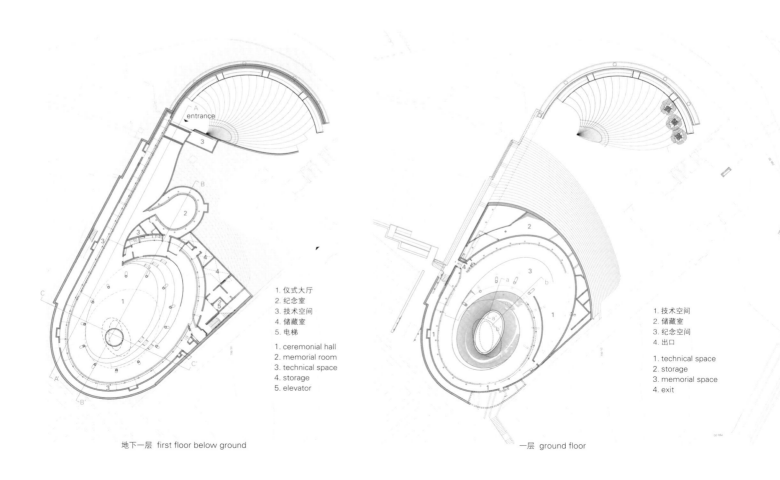

地下一层 first floor below ground

1. 仪式大厅
2. 纪念室
3. 技术空间
4. 储藏室
5. 电梯

1. ceremonial hall
2. memorial room
3. technical space
4. storage
5. elevator

一层 ground floor

1. 技术空间
2. 储藏室
3. 纪念空间
4. 出口

1. technical space
2. storage
3. memorial space
4. exit

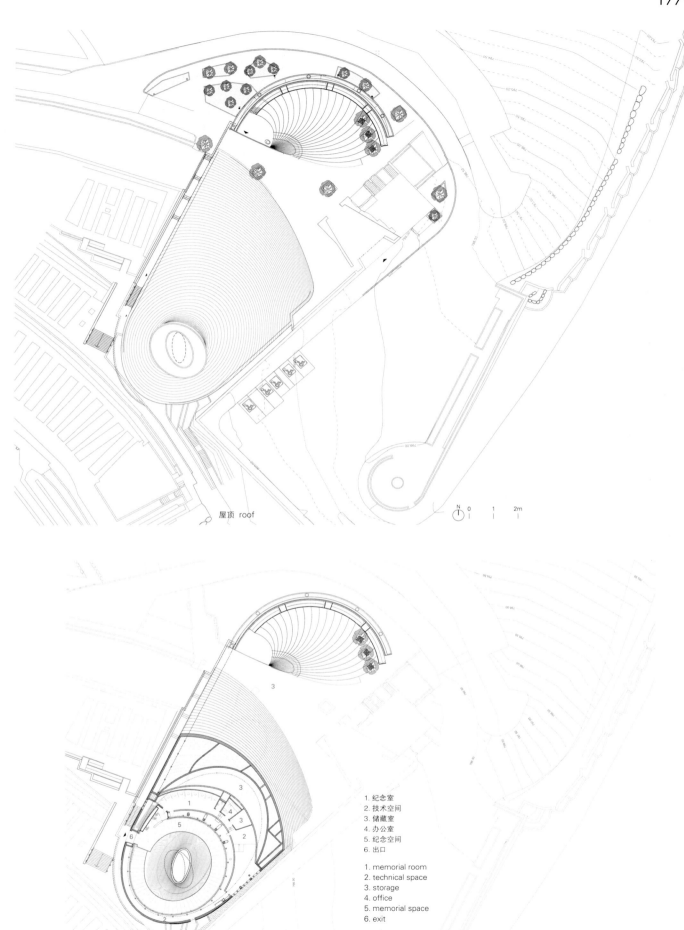

屋顶 roof

二层 first floor

1. 纪念室
2. 技术空间
3. 储藏室
4. 办公室
5. 纪念空间
6. 出口

1. memorial room
2. technical space
3. storage
4. office
5. memorial space
6. exit

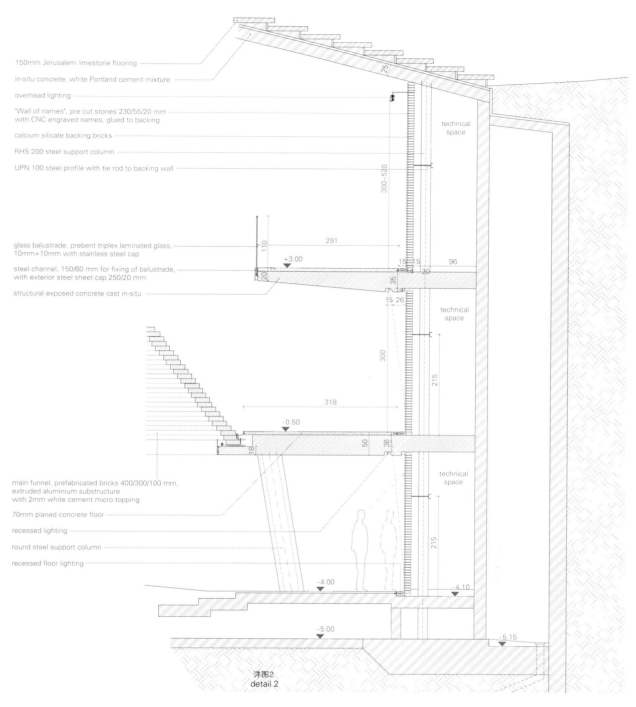

详图2
detail 2

a-a'详图 detail a-a'

b-b'详图 detail b-b'

1. 纪念空间
2. 储藏室
3. 技术空间

1. memorial space
2. storage
3. technical space

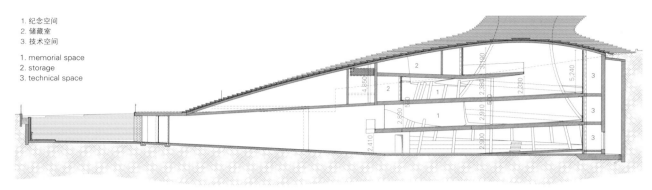

A-A'剖面图 section A-A'

1. 仪式大厅
2. 纪念室
3. 纪念空间
4. 储藏室
5. 技术空间

1. ceremonial hall
2. memorial room
3. memorial space
4. storage
5. technical space

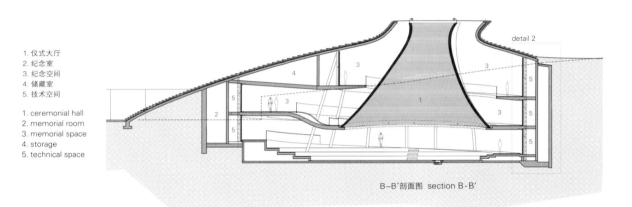

B-B'剖面图 section B-B'

1. 仪式大厅
2. 纪念空间
3. 技术空间

1. ceremonial hall
2. memorial space
3. technical space

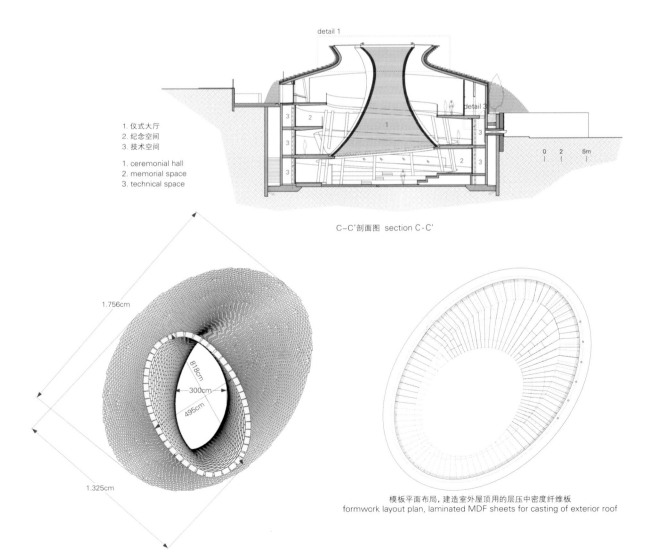

C-C'剖面图 section C-C'

模板平面布局，建造室外屋顶用的层压中密度纤维板
formwork layout plan, laminated MDF sheets for casting of exterior roof

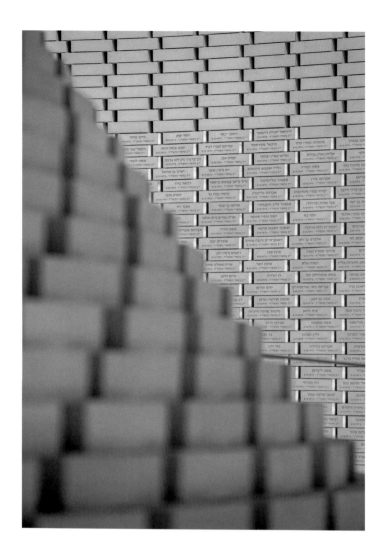

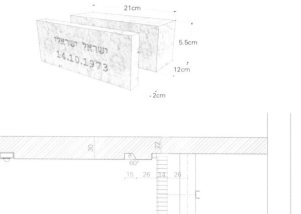

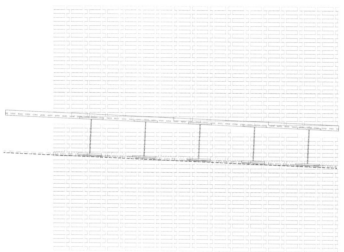

详图3 detail 3

姓名墙立面
Wall of Names elevation

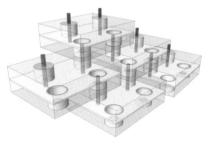

砖金属筒
brick metal cup

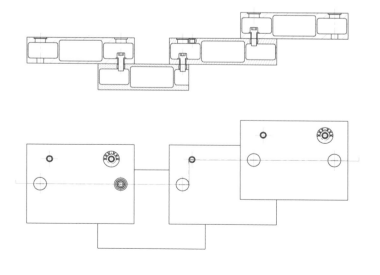

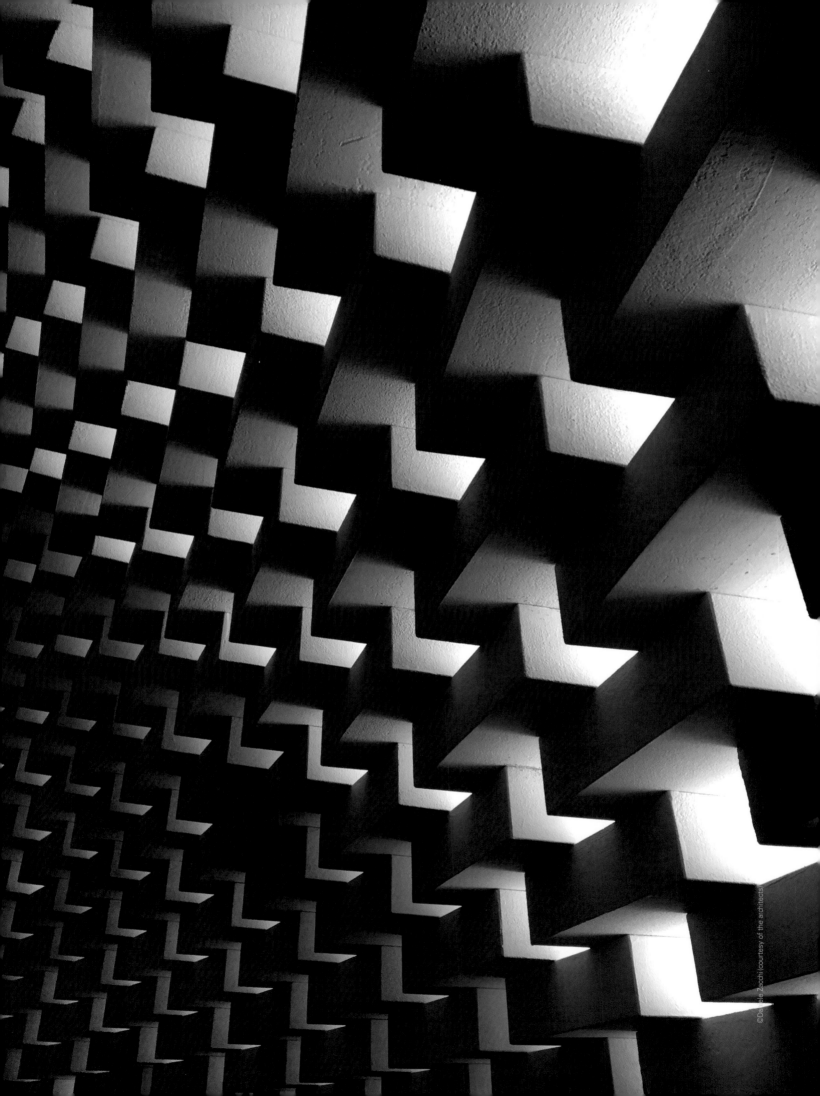

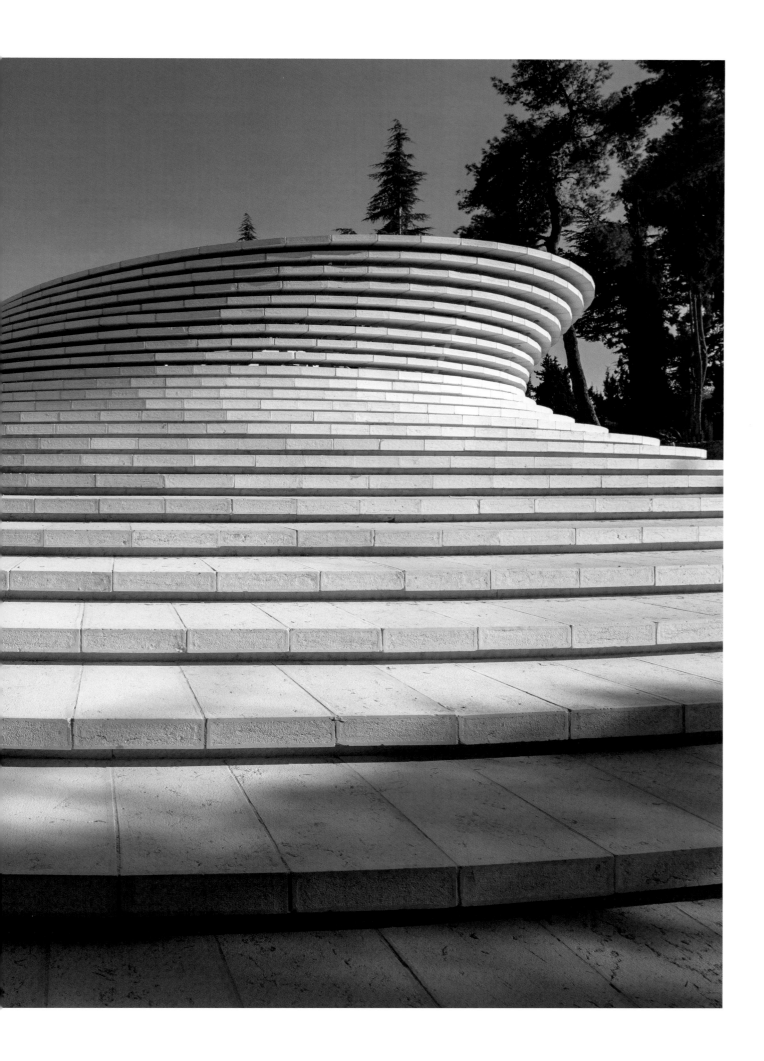

186

渥太华国家大屠杀纪念碑
National Holocaust Monument Ottawa

Studio Libeskind

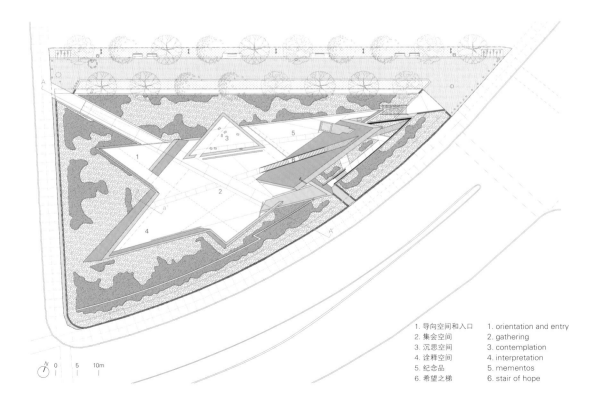

1. 导向空间和入口	1. orientation and entry
2. 集会空间	2. gathering
3. 沉思空间	3. contemplation
4. 诠释空间	4. interpretation
5. 纪念品	5. mementos
6. 希望之梯	6. stair of hope

宏伟庄严的国家大屠杀纪念碑占地3200m²，位于加拿大渥太华市历史悠久的LeBreton Flats地区，惠灵顿大街和布斯大街的交叉处，加拿大战争博物馆对面。

这座纪念碑不仅是为纪念大屠杀中的罹难者而建造的重要公共空间，它还时刻提醒着我们，当今世界正受到反犹太主义、种族主义和偏执主义的威胁。加拿大拥护基本的民主价值观，不论种族、阶级和宗教信仰，而国家纪念碑正是对这些原则和对未来的表达。

现场浇筑的混凝土纪念碑被构思为由六个三角形混凝土体量构成，形成六角星的形状。六角星成为大屠杀的符号象征：六角星一直是大屠杀纪念碑的视觉符号（纳粹迫使犹太人佩戴六角星符号，以便于对犹太人进行识别、隔离和赶尽杀绝）。三角形的空间则代表着纳粹及其同党们给同性恋者、Roma-Sinti种族、耶和华见证会人和政治宗教囚犯钉上的谋杀标签。它由两种意义不同的物理平面组成，上升平面指向未来，下降平面指引游客进入内部空间以尽情沉思和悼念。六个三角形混凝土体量在纪念碑中分别承担不同的功能：加拿大大屠杀历史的诠释空间，三个独立的沉思空间，一个大型中央集会和导向空间以及高耸的、以永恒纪念之火焰为特色的Sky Void。这是一个高达14m的围合体量，它以教堂般的空间包围游客，并框入了上方的一小块天空。

Edward Burtynsky拍摄的巨幅大屠杀遗址的黑白摄影——死亡集中营、杀戮场和森林，精准地绘制在每一面三角形的混凝土墙壁上。这些发人深省的壁画旨在把游客引入由斜墙和迷宫一样的走廊组成的室内空间。

"希望之梯"从中间的集会空间升起，穿透了一面倾斜角度非常大的墙壁，指向上方广场和议会大厦所在方向，寓意承认和答谢那些为加拿大做出重大贡献和在揭露有国家支持的种族灭绝的危险性上持续发挥重要作用的加拿大的幸存者们。

纪念碑四周将会有不同种类的松树从铺满砾石的地面中生长出来。这一景观将一天天变化，更能代表幸存者和他们的子孙为加拿大做出的贡献。

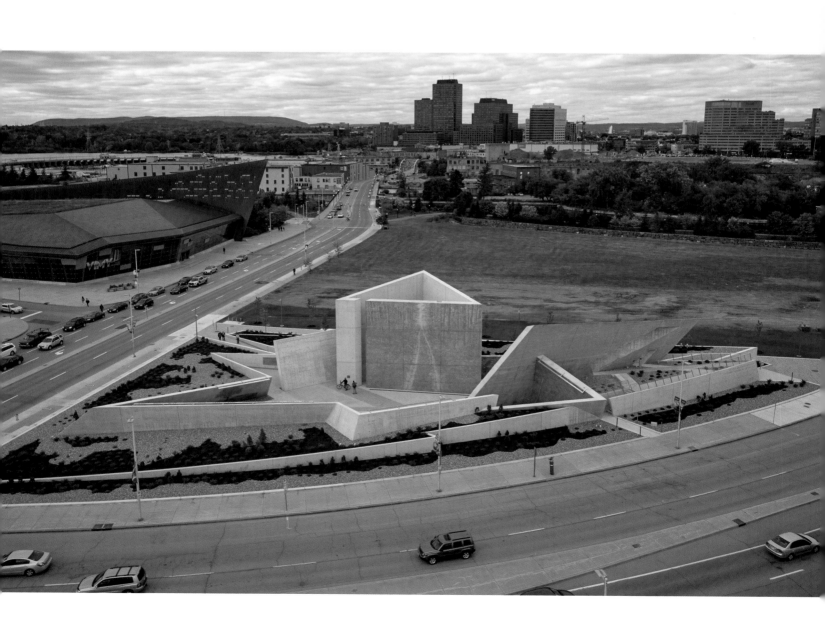

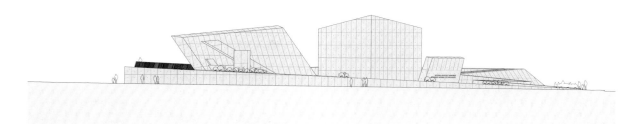

北立面 north elevation

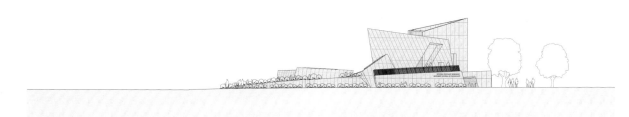

东立面 east elevation

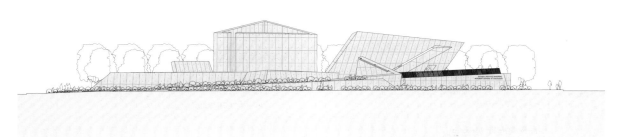

南立面 south elevation

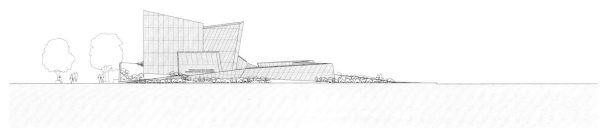

西立面 west elevation

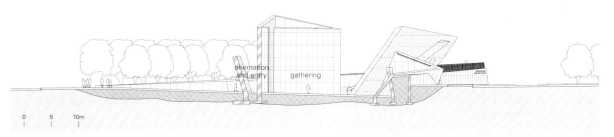

A-A'剖面图 section A-A'

The National Holocaust Monument stands on a 3,200m² site at the intersection of Wellington and Booth Streets within the historic LeBreton Flats in Ottawa, symbolically located across from the Canadian War Museum.

This monument not only creates a very important public space for the remembrance of those who were murdered in the Holocaust, but also serves as a constant reminder that today's world is threatened by anti-Semitism, racism and bigotry. Canada has upheld the fundamental democratic values of people regardless of race, class or creed, and this national monument is the expression of those principles and of the future.

The cast-in-place, exposed concrete monument is conceived as an experiential environment comprised of six triangular, concrete volumes configured to create the points of a star. The star remains the visual symbol of the Holocaust – a symbol that millions of Jews were forced to wear by the Nazi's to identify them as Jews, exclude them from humanity and mark them for extermination. The triangular spaces are representative of the badges the Nazi's and their collaborators used to label homosexuals, Roma-Sinti, Jehovah's Witnesses and political and religious prisoners for murder.

It is organized with two physical ground planes that are differentiated by meaning: the ascending plane that points to the future; and the descending plane that leads visitors to the interior spaces that are dedicated to contemplation and memory. Six triangular concrete forms provide specific program areas within the monument: the interpretation space that features the Canadian history of the Holocaust; three individual contemplation spaces; a large central gathering and orientation space; and the towering Sky Void that features the eternal Flame of Remembrance, a 14-meter-high form that encloses visitors in a cathedral-like space and frames the sky from above.

Edward Burtynsky's large scale monochromatic photographic landscapes of Holocaust sites – death camps, killing fields and forests – are painted with exacting detail on the concrete walls of each of the triangular spaces. These evocative murals aim to transport visitors and create another dimensionality to the interior spaces of canted walls and labyrinth-like corridors.

The Stair of Hope rises from the central gathering space, cuts through a dramatically inclined wall and points at the upper plaza towards the Parliament Buildings, a gesture that recognizes and acknowledges the Canadian survivors who contributed much to Canada and who continued to play an important role in exposing the dangers of state sponsored genocide.

Surrounding the monument, a rough landscape of various coniferous trees will emerge from the rocky pebbled ground. This landscape will evolve over time and it is the representative of how Canadian survivors and their children have contributed to Canada.

项目名称：National Holocaust Monument Ottawas
地点：Ottawa, Canada
建筑师：Studio Libeskind
景观建筑师：Claude Cormier + Associés
墙壁摄影：Edward Burtynsky
结构工程师：Read Jones Christofferson
照明设计：Focus Lighting
机电土木工程师：WSP
总承包商：UCC Group
主要文化顾问：Lord Cultural Resources
客户：Government of Canada National Capital Commission
总楼面面积：3,180m²
最大高度（Sky Void）14m from exterior elevation,
15m from local interior floor level
材料：Structure – exposed, smooth, cast-in-place concrete;
Walls – exposed, smooth, cast-in-place concrete; Balustrades and
Guardrails – galvanized steel;
Floors – cast-in-place concrete with light sand blast finish
and exposed aggregate (accents);
Stair of Hope – cast-in-place concrete
竣工时间：2017.9
摄影师：©Doublespace (courtesy of the architect)

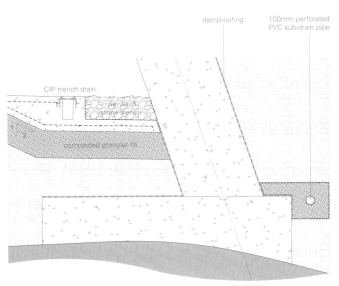

a-a'详图 detail a-a'

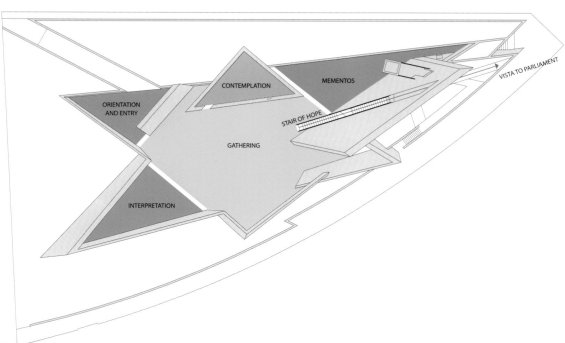

第二次世界大战博物馆
The Second World War Museum

Studio Architektoniczne Kwadrat

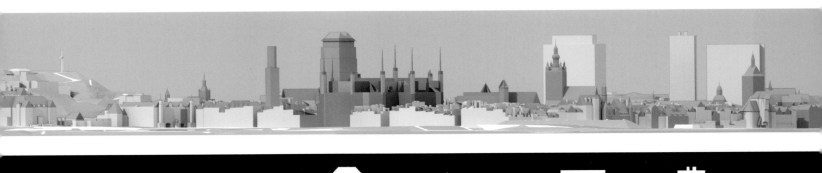

The Second World War Museum

40.0m

第二次世界大战博物馆坐落在一个具有象征意义的建筑空间中，也是一个记忆空间，距离格但斯克市历史悠久的波兰邮局200m，距离河对岸的韦斯特普拉特半岛3km。这两处都曾在1939年9月遭遇袭击。

这个为博物馆开辟的1700m²的场地在西面与Radunia运河相接，南面面向莫特拉瓦河的广阔景观。这些地方如今都属于老格但斯克市的郊区，但是不久它将取代船坞，成为这座城市的现代化核心区域。为了适应这座城市的历史元素，此建筑的设计将会是对第二次世界大战悲剧的微妙暗示而非强烈引证。

这座博物馆描绘出了以往的悲剧和当下的活力，为未来开阔了视野。这座21世纪的博物馆建筑把传统城市空间、规模、材料和城市色彩紧密结合在一起，与格但斯克市的标志性天际线相呼应。新公共广场上方耸立着一座高40.5m的塔，具有与地面形成45°角的表面，从不同方向观察时外观都不一样，且富有动态感。四个梯形立面中的三个都被红色陶瓦板覆盖，而第四个面和屋顶由玻璃覆盖，这样就可以让自然光涌入室内。简单的雕塑形式，完全没有任何字面含义，能引发各种联想。它曾被比作堡垒、屏障、濒临倒塌的房屋或是碉堡，而在夜晚被照亮时，它就像一支燃烧的蜡烛。

这座博物馆三个空间分区象征着战时、现在和未来的关系：过去隐藏在建筑地下层，现在呈现于建筑周围的开阔空间，未来通过该建筑的上升突起部分表达，包括一个观景平台。

进入地下层将会是一个心情不断转换的过程。从一开始的不以为然和充满日常想法到思维清晰，最后陷入恐惧，甚至为与展览之间的深刻关系而感到痛楚。参观者将从博物馆地下穿过战争地狱的小路开始他们的旅程，并在结束时看到塔顶的平台。

这座第二次世界大战博物馆不仅会成为一个独特而有力的图标，还会成为一个代表着永不遗忘的第二次世界大战历史的公共景点和一个铭刻在格但斯保市、波兰和欧洲人民心中的新标志。

The Second World War Museum is located in a symbolic architectural space, which is also a space of memory, 200 meters from the historic Polish Post Office in Gdańsk and 3 kilometers across the water from Westerplatte Peninsula, both of which were attacked in September 1939.

The 1,700m² lot set aside for the museum touches the Radunia Canal to the west, while its south side opens onto a wide panorama of the Motława River. Today, these are the outskirts of Old Gdańsk but, soon, it will be the core of the modern section of the city that will replace its shipyard. To fit in the historic part of the city, the building was designed to be a delicate suggestion rather than strong quotation for the Second World War tragedy.

The museum narrates the tragedy of the past, the vitality of the present, and opens the horizons of the future. The rising, dynamic form symbolizes the museum below, while giving a panoramic and spectacular orientation to the historic city and its future. Echoing the iconic skyline of Gdańsk, the building ties together traditional urban spaces, scales, materials, and colours of the city with a 21st-century museum.

A tower rising 40.5 meters above the new public square features surfaces set at angles as much as 45 degrees from the vertical, lending it a dynamic appearance that alters

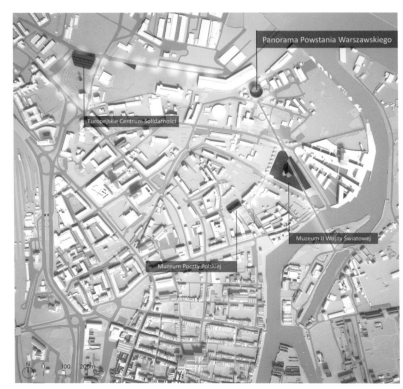

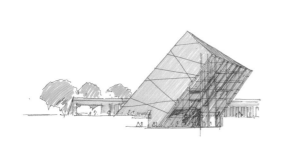
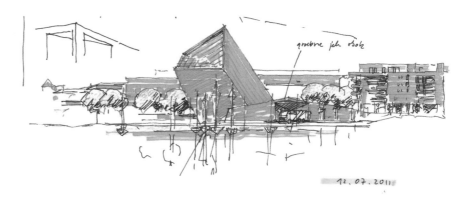

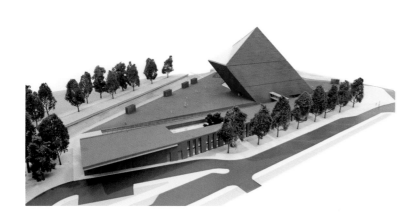

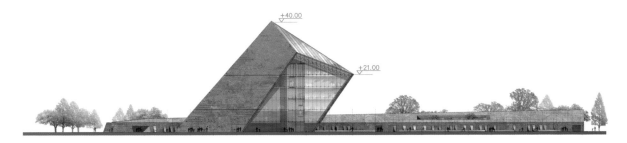

西南立面 south-west elevation

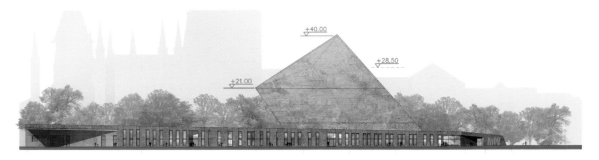

东北立面 north-east elevation

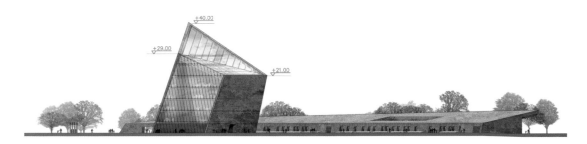

南立面 south elevation

东南立面 south-east elevation

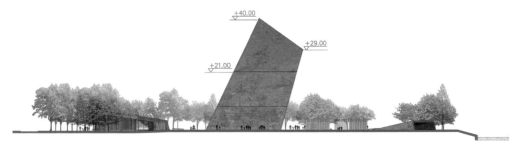

西北立面 north-west elevation

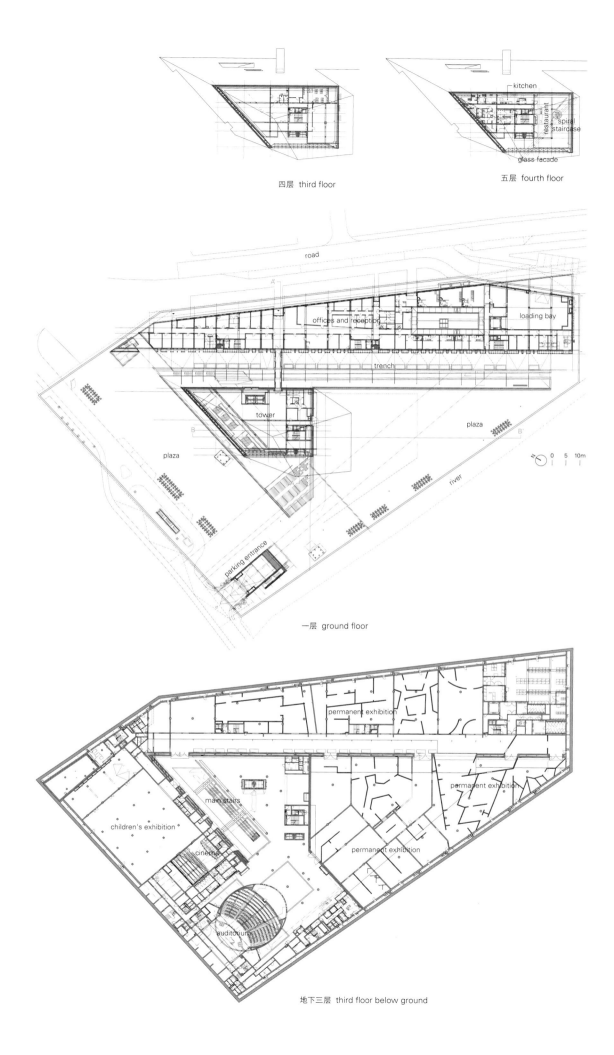

when viewed from different directions. Three of its four trapezoidal facades are clad in terracotta-red panels, while the fourth side and kinked roof are filled in with glazing that allows natural light to flood into the interior. This simple sculptural form, devoid of literal meaning, evokes various associations. It has already been likened to a bastion, a barrier, a crumbling house or a bunker, and when illuminated at night, it resembles a burning candle.

The museum's spatial division into three areas symbolizes the relationships between the wartime past, the present and the future: the past is hidden on the building's underground levels, the present appears in the open space around the building and the future is expressed by its rising protrusion, which includes a viewing platform.

Entering the subterranean levels is to be a mood setting process. Starting from being unconcern and full of everyday thoughts, to be hanged in the balance and clear minded, to finally fell the horror, frightens and even pain by a strong relation with the exhibition. The visitors may begin their journey from underground of the museum, a path through hell of war, and finish their time travel experience at the observation deck on the top of the tower.

This Second World War Museum will become a unique and powerful icon, as well as a public attraction standing for the never-to-be-forgotten history of the Second World War, a new symbol that will inscribe itself in the hearts of the people of Gdańsk, Poland and Europe.

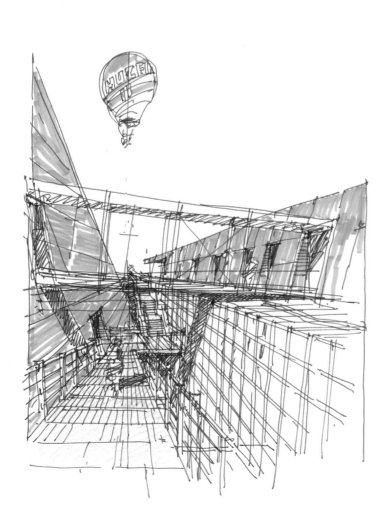
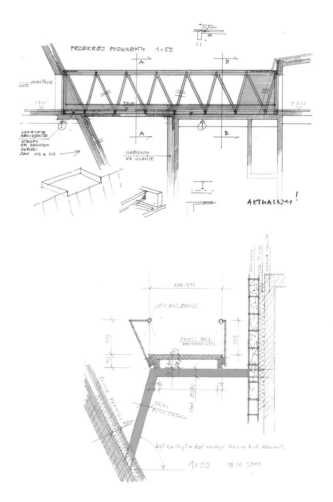

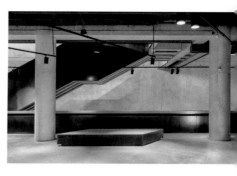
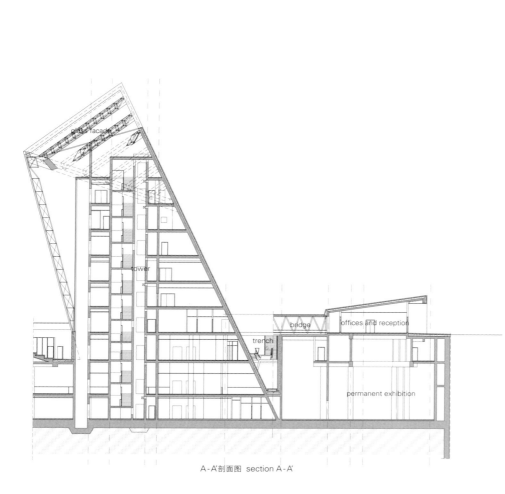

A-A'剖面图 section A-A'

B-B'剖面图 section B-B'

项目名称：The Second World War Museum / 地点：Na Dylach, Stara Stocznia, ulice Wałowa, Gdańsk, Poland
建筑师：Studio Architektoniczne Kwadrat–Jacek Droszcz, Bazyli Domsta, Andrzej Kwieciński, Zbigniew Kowalewski
室内设计：LOFT Magdalena Adamus / 景观建筑师：Studio Architektoniczne Kwadrat / 施工：PG PROJEKT / 总承包商 WARBUD S.A., Hochtief Polska S.A., Hochtief Solutions AG / 客户：Muzeum II Wojny Światowej W Gdańsku / 造价：450 mln PLN / 用地面积：17,095m²
建筑基底面积：4,465.40m² / 可使用面积：36,058.24m² / 总面积：57,386m² / 竞赛时间：2010 / 设计时间：2011—2012 / 施工时间：2012—2017
摄影师：©Tomasz Kurek (courtesy of the architect)

©Stefan Ruiz

P24 **ALA Architects**
Specializes in demanding cultural buildings, terminal design and unique renovation projects. The Helsinki–based firm was founded in 2005 by four partners: Juho Grönholm, Antti Nousjoki, Janne Teräsvirta and Samuli Woolston. Their collaboration started in 2004 through success in architectural competitions. Received the 1st prize in the open international competition for the new theater and concert hall, Kilden Performing Arts Centre, in Kristiansand, Norway in 2005. Is today run by Grönholm, Nousjoki and Woolston, and in addition to them employs 38 architects, interior designers, students and staff members, representing 14 nationalities. All three partners have around 20 years of professional experience, mostly in designing large public buildings both in Finland and abroad. In 2012 they received the prestigious Finnish State Prize for Architecture.

P48 **De Smet Vermeulen Architecten**
Founded in 1989, persues the architecture which is not specialized but also is generalized. Aims to work in a multitude of fields, ranging from interior to urban design and policy support. Works with collaborators to complement their ability. Tries to look beyond the project, thinking architecture is urbanism created from the bottom up. Also approaches the architecture from the long term perspective to persue the sustainability of building. Henk De Smet teaches at KU Leuven, heading a studio on sustainable architecture. Paul Vermeulen is an author who has written extensively about architecture. He won the Flemisch Culture Prize for architecture in 2011.

P186 **Studio Libeskind**
Was established by Polish-American architect, Daniel and his partner Nina Libeskind in Berlin, Germany in 1989, after winning the competition to build the Jewish Museum Berlin. In 2003, the studio was selected to develop the master plan for the World Trade Center and moved its headquarters from Berlin to New York City to oversee the project. It has offices in New York and Zurich. Creates architecture that is resonant, practical and sustainable. The studio's completed buildings range from museums and concert halls to convention centers, university buildings, hotels, shopping centers and residential towers.

P200 **Studio Architektoniczne Kwadrat**
Was established in 1989 in Gdynia, Poland, by Jerzy Kaczorowski, Jacek Droszcz and Adam Drohomirecki, leaders of the project group of Department of Architecture of Gdansk Institute of Technology. Has been acclaimed after winning several awards in architecture and urban competition. Since 1995, the studio has been led by Jacek Droszcz and Bazyli Domsta. Both work as architect and urban planner, as well as lecturer at the Faculty of Architecture in the Gdansk University of Technology. Jacek Droszcz has designed several architectural and urban complexes, such as offices, shopping centers, museums, churches and homes and residential areas. He is the winner of 42 architectural contests.
Studio has performed several projects including residential architecture and public utilities. Potential of the office has been based on many years experiences of designers and creativeness of young architects and students. Has lead architects, but every architect in the office is developing individuality and creative passion here. Has participated in numerous architectural competitions and received more than 40 awards and honors.

P14 **baukuh**
Was founded in 2004. Based in Milan and Genoa, it is now composed by Paolo Carpi, Silvia Lupi, Vittorio Pizzigoni, Giacomo Summa, Pier Paolo Tamburelli and Andrea Zanderigo. Participated in several biennales, including the Rotterdam (2007 and 2011), in the Istanbul (2012) in the Venice (2008 and 2012) and was part of the research group for the Dutch National History Museum (2011). Vittorio Pizzigoni teaches at the University of Genoa, Pier Paolo Tamburelli teaches at University of Illinois at Chicago, Andrea Zanderigo is a teaching assistant at EPFL Lausanne.

P58 Mindspace

Founding partner, Sanjay Mohe graduated from the Sir JJ College of Architecture in Mumbai and established "Mindspace" in Bengaluru, India. His work spans a spectrum of projects - research laboratories, knowledge parks, campus designs, factories, beach resorts, libraries, corporate offices, hospice and residences. Some of the award winning projects include: The Golden Architect Award by A+D & Spectrum Foundation Architecture Award (2009); J K Cements Architect of the Year Award (seven times since 1991); The Award of the Journal of the Indian Institute of Architects (2002); ar+d International Annual Award of *The Architectural Review* (London) + *d line* for the JRD Digital Library Bangalore (1999); Gold Medal from the Asian Forum for Institutes of Architecture (1998).

P140 BBGK Architekci

Was established by Jan Belina Brzozowski[left], Konrad Grabowiecki[center] and Wojtek Kotecki[right]. Based in Warsaw, the office is comprised of over 30 architects. Has extensive experience in the field of historic, public, residential buildings and urban design. They design individual architecture strongly related to the existing context, as evidenced by the Katyń Museum put into use in 2015. Was awarded in many competitions and selected as finalists for the 2017 EU Prize for Contemporary Architecture – Mies van der Rohe Award. Their work addresses the changing role of the future cities.

©Bartek Barczyk

P154 Architectural Design & Research Institute of SCUT South China University of Technology

Is a well-known outstanding design and research institute in China. Has established itself as a brand for designing cultural, education, mega-high-rise, sports and exhibition architecture.

He Jingtang, architect, holds the position of the honorary dean of the School of Architecture of South China University of Technology and the director of the Architectural Design and Research Institute of South China University of Technology. Ni yang, He Chili, He Xiaoxin, Liu Tao are the key members of SCUT. Professor He Jingtang, designed a number of campus buildings, including Zijingang Campus of Zhejiang University and Hengqin Campus of Macau University. He also designed influential public buildings include 2010 Shanghai World Expo China Pavilion, the expansion project for Memorial Hall of the Victims in Nanjing Massacre by Japanese Invaders, Tianjing museum, Yingxiu Epicenter Memorial Hall, Qian Xuesen Memorial Hall, Museum of Nanyue King Tomb of the Western Han Dynasty, and etc.

P48 Studio Roma

Is a Belgium architecture studio founded in 2010 by Piet Stevens, Sofie Beyen and Marc Vanderauwera. Is now consisted of four partners, including newly joined Sarah Vaelen in 2015. With the years of experience in the architectural restoration, they have been actively participated in research and completion of re-using architectural heritage. Its work also includes master plan, redevelopment projects and architectural policy proposal to the city.

P168 Kimmel Eshkolot Architects

Was founded in Tel Aviv, Israel in 1986 by Etan Kimmel[left] and Michal Kimmel Eshkolot[right]. They graduated from the Faculty of Architecture and Town Planning, Technion-Israel Institute of Technology, Haifa in 1985. In their first years of practice, they were involved in the preservation and rehabilitation of Neve Tzedek, the historical neighborhood of Tel Aviv. In 1993 they were awarded the Rokach prize for their projects in Neve Tzedek. Won the Rechter Prize for Architecture, considered to be the most prestigious award for architecture in Israel, 2011. The practice is currently involved in dozens of projects in different scales, both in Israel and Europe.

Olga Sezneva

Is an urban sociologist at the University of Amsterdam. Received her Ph.D. from New York University and taught as a Harper & Schmidt Fellow at the University of Chicago. Olga's interest in cities leads her to research and write on the issues of architecture and social memory, public space and state politics, migration and displacement, and the ways in which different cultures encounter each other in the city. Her work appeared in journals *Environment and Planning D, International Journal of Urban History and Critical Historical Studies*. Olga is a member of the Board of Experts of the European Prize for Public Space. Curates an international collective or artists and scientists Moving Matters Traveling Workshop, which explores issues of migrations through performance, visual art and literary writing.

Diego Terna

Received a degree in architecture from the Politecnico di Milano and has worked for Stefano Boeri and Italo Rota. Has been working as a critic and collaborating with several international magazines and webzines as an editor of architecture sections. In 2012, he founded an architectural office, Quinzii Terna together with his partner Chiara Quinzii. Currently is an assistant professor of Politecnico di Milano and runs his personal blog *L'architettura immaginata* (diegoterna.wordpress.com).

P70 BCHO Architects

Since Cho Byoung-soo opened his office in 1994, he has been actively practicing with themes such as 'experience and perception', 'existing and existed', 'I shaped house, L shaped house', 'contemporary vernacular' and 'organic versus abstract'. Has taught the theory and design of architecture at several universities including Harvard University, Columbia University and University of Hawaii. Is the recipient of Architizer A+ Awards and AIA Northwest and Pacific Region Design Awards in 2015, KIA Award in 2014 with several previous AIA Honor Awards in Montana Chapter and in N.W.Pacific Regional. His projects have been published in various magazines including the *Architectural Reveiw*, *Dwell*, and *deutsche bauzeitung(db)*.

P36 A. Lerman Architects

Asaf Lerman(1969) studied architecture from 1994 to 2000 at the AA School of Architecture in London and received ARB/RIBA Part 2. Founded AleCC (Asaf Lerman & Celine Condorelli Architecture) in London, 2001 and A. Lerman Architects ltd. in Tel Aviv, 2006. Has taught at the Betzalel academy for arts and architecture in Jerusalem from 2003 to 2017. Today the firm focuses on a diverse set of projects ranging from public works to tech, hotels, commercial, and residential. Its offices have a history of collaboration with leading international architects on projects in Israel, like the Jerusalem National Museum, The Mandel School at the Hebrew University, and the Bezalel School for the Arts.

P132 Nizio Design International

Was founded in 2002 by Miroslaw Nizio[p.228-lower] after his initial design business in the 1990s in New York. He studied at the Department of Interior Architecture and Sculpture, Academy of Fine Arts in Warsaw, and at the faculty of Interior Design, Fashion Institute of Technology in New York. He is recognized internationally for designs of public spaces including museums, historical exhibitions and commemoration monuments. The office has long-standing experience of designing and completing architectural and revitalization projects. The exhibitions designed by Nizio reconstruct history balancing the latest technologies and artistic means of exhibition. Major projects are the Mausoleum of the Martyrdom of Polish Villages in Michniów and revitalization of the former synagogue in Chmielnik into the home of 'Świętokrzyski Shtetl', the educational and museum facility featuring a studio-designed exhibition.

P120 **SANDWICH**

Kohei Nawa[p.229-lower, right], director of SANDWICH, is a Sculptor as well as a Professor of the Kyoto University of Art and Design. He was born in Osaka, 1975. Received a Bachelor of Fine Arts in Sculpture in 1998, a Master of Fine Arts in sculpture in 2000 and a PhD in Fine Arts and Sculpture in 2003 from the Kyoto City University of Arts. Established SANDWICH in Kyoto in 2009.

P84 **Pattersons**

Founder and Director, Andrew Patterson is an Auckland, New Zealand-based architect. Born in the Waikato region in 1960, he completed a Bachelor of Architecture degree at The University of Auckland in 1984 and started his own practice at age 26. Received Distinguished Alumni award in 2013 from the Auckland University. Is currently teaching at Unitec as a guest professor. Is a Fellow of the New Zealand Institute of Architects[FNZIA]. Also is a member of the Auckland and Queenstown Urban Design Panels and President of the Auckland Architecture Association. During his career he has won New Zealand National Awards for Architecture five times. In 2017 Patterson was awarded the FNZIA's Gold Medal.

P104 **Blakstad Haffner Architects**

Erlend Blakstad Haffner is a Norwegian architect and landscape architect. Graduated from the Bergen School of Architecture (Bachelor in Architecture) and The Bartlett, UCL (Diploma affiliate in Architecture) and The CASS, London Metropolitan University (Master of Architecture). Founded Fantastic Norway AS in 2004 and was until 2014 head of the office. In 2014, Fantastic Norway was discontinued (now Tiund AS) and he started Blakstad Haffner Architects. His architectural experience spans over a broad range; from the preparation of broad sketches and concepts, preliminary projects and detailed projects, to follow-ups on construction sites where dialogue with the building contractors is an important factor for success. Has taught at the architectural schools in Bergen, Oslo, London, Sao Paulo and Cornell, NY.

P120

© 2019大连理工大学出版社

版权所有·侵权必究

图书在版编目(CIP)数据

记忆空间：汉英对照 / 德国里伯斯金建筑事务所等编；于风军等译. — 大连：大连理工大学出版社，2019.4

（建筑立场系列丛书）

ISBN 978-7-5685-1955-7

Ⅰ．①记… Ⅱ．①德… ②于… Ⅲ．①纪念建筑－研究－汉、英 Ⅳ．①TU251

中国版本图书馆CIP数据核字(2019)第062517号

出版发行：大连理工大学出版社
　　　　　（地址：大连市软件园路80号　邮编：116023）
印　　刷：上海锦良印刷厂有限公司
幅面尺寸：225mm×300mm
印　　张：13.75
出版时间：2019年4月第1版
印刷时间：2019年4月第1次印刷
出 版 人：金英伟
统　　筹：房　磊
责任编辑：杨　丹
封面设计：王志峰
责任校对：张昕焱
书　　号：978-7-5685-1955-7
定　　价：258.00元

发　　行：0411-84708842
传　　真：0411-84701466
E-mail：12282980@qq.com
URL：http://dutp.dlut.edu.cn

本书如有印装质量问题，请与我社发行部联系更换。

建筑立场系列丛书 01：
墙体设计
ISBN: 978-7-5611-6353-5
定价：150.00元

建筑立场系列丛书 09：
墙体与外立面
ISBN: 978-7-5611-6641-3
定价：180.00元

建筑立场系列丛书 17：
旧厂房的空间蜕变
ISBN: 978-7-5611-7093-9
定价：180.00元

建筑立场系列丛书 25：
在城市中转换
ISBN: 978-7-5611-7737-2
定价：228.00元

建筑立场系列丛书 33：
本土现代化
ISBN: 978-7-5611-8380-9
定价：228.00元

建筑立场系列丛书 41：
都市与社区
ISBN: 978-7-5611-9365-5
定价：228.00元